Contents

Acknowledgements

I wish to thank William Foster for allowing me to use the collection and library of the insect room at the Cambridge University Museum of Zoology. Thanks also to Henry Disney for valuable advice and the use of facilities at Malham Tarn Field Centre, to Steve Tilling for his help and guidance through the testing stage and to Laurie Barnes for her comments on the final manuscript. And finally, thanks to all those kind people who helped in the testing of the key.

This AIDGAP key was first published in *Field Studies* (6): 149-197 in 1984, and then reprinted with minor alterations in 1988. It has been reprinted with a new cover and revised page numbering in 2020.

Introduction

The vast majority of beetles (order Coleoptera) can be distinguished from other adult insects by the following two characters:

i) The forewings are reduced to hard or leathery elytra (wing cases) which usually meet on the centre-line of the abdomen.

ii) They have biting (rather than sucking) mouthparts.

Beetles are unlikely to be confused with any other order, except the Hemiptera (true bugs). There are some beetles with elytra reduced or even completely absent. These can look superficially like immature bugs, although bugs have a rearward-directed rostrum through which they suck their food. The characters that distinguish beetles from bugs are shown in Fig. 1.

The order Strepsiptera consists of three families of unusual insects that parasitise Hymenoptera and Hemiptera. The females are not free-living, but the males are distinguished by having fully-developed hindwings and reduced forewings (the opposite way round to flies). They are included in this beetle key because some authorities believe that they really belong to the Coleoptera. However, one has to adopt a standard of nomenclature for a key, and in the standard adopted for this one (Pope, 1977) the Strepsiptera are taken as a separate order. In order to examine beetles it is necessary to kill them. The easiest way is probably to use a killing jar. This can be made by casting a layer of plaster of Paris at the bottom of a glass jar. The plaster will then absorb a liquid killing agent, such as ethyl acetate. To avoid degradation of the specimen, only the vapour should come into contact with it. Beetles may also be killed by placing them in a tube of alcohol.

Beetles may be kept in alcohol, pinned or mounted on cards. Ecologists will often wish to keep their specimens in alcohol, but others wishing to build up a collection will usually prefer some dry form of preservation. Pinning is quicker than card mounting, and allows the underside of the beetles to be examined with ease. The pin should pass through one of the elytra, about a third of the way from the base, and just outboard of the centre-line. Card mounting makes the underside more difficult to examine, but is very tidy and gives the specimen a good deal of protection. For card mounting, a gum adhesive is required. Commercial gums such as 'gloy', may be used, although they tend to be rather thick and this can obscure details of the tarsi (see Fig. 3). A gum adhesive can be made by taking a small quantity of gum tragacanth (available from chemists) and adding absolute alcohol while stirring, until the volume of mixture is about double that of the original powder. Then, approximately four times this volume of water should be added to make a smooth paste. Details of card mounting techniques are given by Joy (1932) and Walsh and Dibb (1974).

Fig. 2 shows the top view of a beetle. The part labelled 'thorax' is strictly the prothorax; the other two thoracic segments (mesothorax and metathorax) are hidden under the elytra. Fig. 3 shows the underside of one of the click beetles (Elateridae), showing the rearward extension of the underside of the prothorax (the prosternum) which fits into a cavity in the second thoracic segment. This character, which is absent in most beetles, can usually be seen from the side, even in card-mounted specimens.

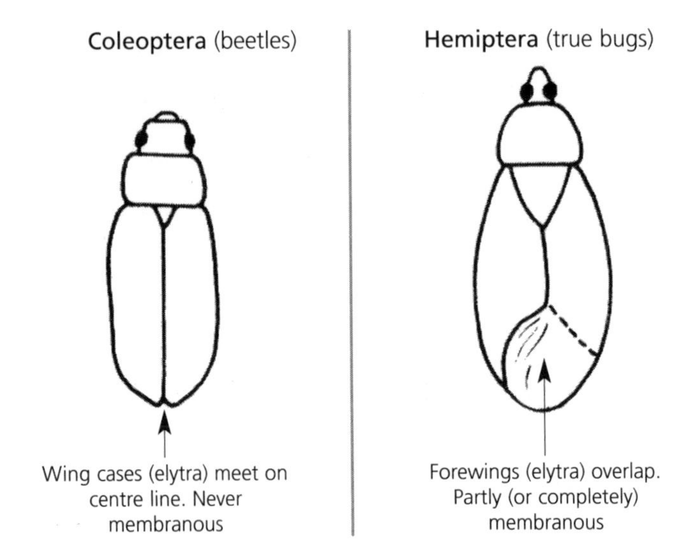

If the elytra are short, not covering most of the abdomen, it is safer to ignore the character above.

Head in profile

Biting mouthparts

Palps

No palps

Rostrum (sucking mouthparts), usually points rearwards

If a beetle has the head extended into a rostrum, it always points forewards, and the beetle has very obvious antennae

Some water bugs have a very short rostrum, but it never points forewards. In this case, antennae hidden

Figure 1. How to distinguish between beetles and bugs.

At the end of this key will be found an alphabetical list of beetle families, with additional information on their characteristics and habitats. Reference is also made here to illustrations in two books on beetles. The first, by Joy (1932), is an extremely useful book although itis now somewhat outdated; some names have been changed, taxonomic 'mistakes' have been rectified and 'new' beetles have been discovered. Nevertheless, used with care, it is one of the most useful entomological texts available. Since it has been reprinted in somewhat reduced format, note that the bars indicating the size of the beetles (beside the illustrations) are about 20% short. The second book, by Lyneborg (1977) is simply a small book of coloured illustrations of beetles. It is considerably less expensive than Joy's book, and used in conjunction with this key will help to illustrate what the beetles actually look like.

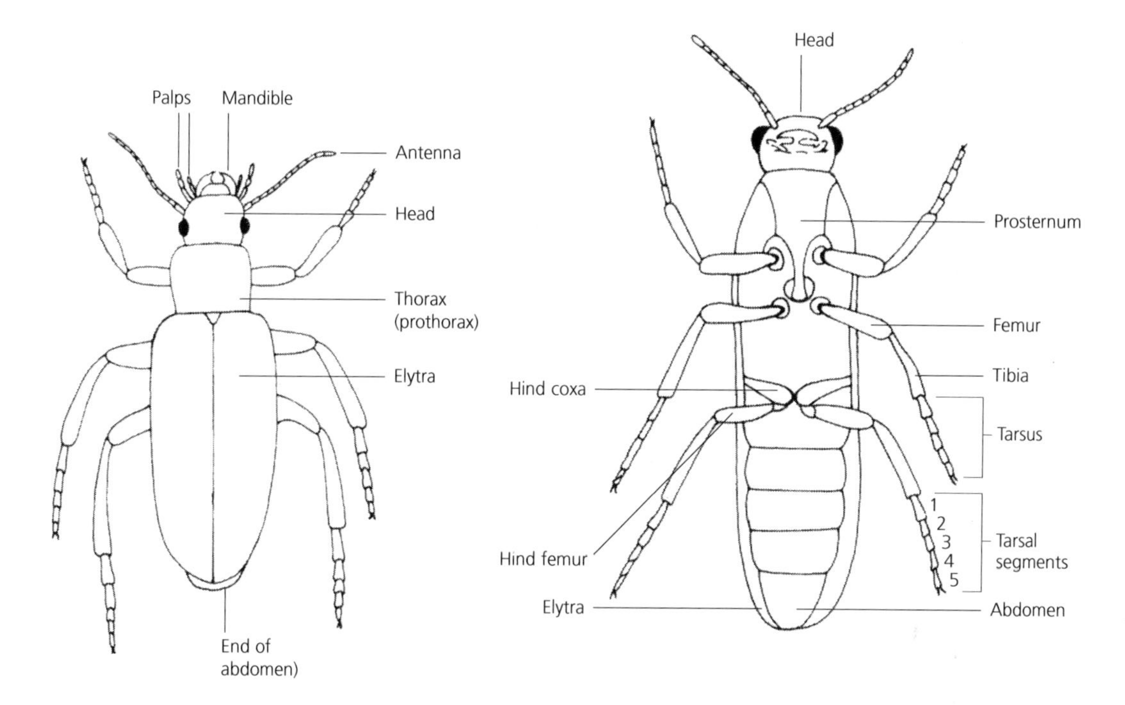

Figure 2. Top (dorsal) view of beetle.

Figure 3. Underside (ventral) view.

To identify beetles beyond the family stage, there are a number of keys available: Joy (1932) keys out most of the British beetles using characters visible from above and from the side, so that card-mounted specimens can be used. More up-to-date is the comprehensive (and expensive) eleven-volume series (in German) on the middle European beetles by Freude *et al.* (1965). Many British beetles are in families which have been the subject of keys in the Royal Entomological Society of London's Handbooks, which are included in the references to this key. Further details, and information on new keys in this series, may be obtained from the Registrar, Royal Entomological Society of London, 41 Queen's Gate, London SW7 SHT. They also publish a check list (Pope, 1977). For information on the biology of the Coleoptera, see Crowson (1981); this is not only a very comprehensive text, but also references many other works of interest. Useful information on all aspects of beetles can be found in the book by Walsh and Dibb (1974), whilst a bibliography of key works is given in Kerrich *et al.* (1978).

Table of families as recognised in this key (after Pope, 1977)

The page numbers in brackets refer to Joy, 1932 (Volume 1). It should be remembered that this excellent book is now quite old, should be used with great care, and always in conjunction with the check list (Pope, 1977).

COLEOPTERA
- **Adephaga**
  - Caraboidea
    - Carabidae (p. 322, Cicindelidae and Carabidae)
    - Haliplidae (p. 259)
    - Hygrobiidae (p. 263, g. *Hygrobia*)
    - Noteridae (p. 272, g. *Noterus*)
    - Dytiscidae (p. 263)
    - Gyrinidae (p. 257)
- **Myxophaga**
  - Sphaeroidea
    - Sphaeriidae (p. 519)
- **Polyphaga**
  - Staphyliniformia
    - Hydrophiloidea
      - Hydrophilidae (p. 281, all the Palpicornia except Hydraenidae)
    - Histeroidea
      - Sphaeritidae (p. 483)
      - Histeridae (p. 468)
    - Staphylinoidea
      - Hydraenidae (p. 295)
      - Ptiliidae (p. 568)
      - Leptinidae (p. 485)
      - Leiodidae (p. 544, as Anisotomidae)
      - Silphidae (p. 462, Necrophoridae and Silphidae)
      - Scydmaenidae (p. 478)
      - Scaphidiidae (p. 475)
      - Staphylinidae (p. 3)
      - Pselaphidae (p. 149, Clavigeridae and Pselaphidae)
  - Scarabaeiformia
    - Scarabaeoidea
      - Lucanidae (p. 242)
      - Trogidae (p. 246, g. *Trox*)
      - Geotrupidae (p. 245, g. *Geotrupes*)
      - Scarabaeidae (p. 243)
  - Dascilliformia
    - Dascilloidea
      - Dascillidae (p. 429)
    - Eucinetoidea
      - Clambidae (p. 482)

Eucinetidae (not included in Joy, 1932)
Scirtidae (p. 424, as Helodidae)
Byrrhoidea
Byrrhidae (p. 483)
Dryopoidea
Psephenidae (p. 425, g. *Eubria*)
Heteroceridae (p. 476)
Limnichidae (p. 484, g. *Limnichus*)
Dryopidae (p. 498, as Parnidae)
Elmidae (p. 4 77, as Helmidae)
Buprestoidea
Buprestidae (p. 450)
Elateroidea
Elateridae (p. 438)
Throscidae (p. 442, g. *Trigaxus*)
Eucnemidae (included in Elateridae, p. 438)
Cantharoidea
Drilidae (p. 427, g. *Drilus*)
Cantharidae (p. 429)
Lampyridae (p. 424)
Lycidae (p. 423)
Bostrichiformia
Dermestoidea
Dermestidae (p. 496)
Bostrichoidea
Anobiidae (p. 456)
Ptinidae (p. 453)
Bostrichidae (p. 459)
Lyctidae (p. 507)
Cucujiformia
Cleroidea
'Clavicornia'
Phloiophilidae (p. 566, g. *Phloiophilus*)
Trogossitidae (p. 516, g. *Nemosoma* p. 487, g. *Tenebroides*)
Peltidae (p. 559, g. *Thymalus*)
Cleridae (p. 427)
Melyridae (included in Cantharidae, p. 429)
Lymexyloidea
Lymexylidae (p. 429, as Lymexylonidae)
Cucujoidea
Nitidulidae (p. 528, Cateretidae and Nitidulidae)
Rhizophagidae (p. 517, g. *Rhizophagus* p. 499, g. *Monotoma*)
Sphindidae (p. 554, g. *Sphindus* p. 559, g. *Aspidiphorus*)
Hypocopridae (p. 509, g. *Hypocoprus*)
Cucujidae (p. 486)
Silvanidae (p. 501, most: p. 486, g. *Psammoecus*)

Cryptophagidae (p. 500)
Biphyllidae (p. 531, as g. *Diphyllus* p. 502, g. *Diplocoelus*)
Byturidae (p. 559, g. *Byturus*)
Erotylidae (p. 530, as Erotylinae)
Phalacridae (p. 526)
Cerylonidae (p. 516, g. *Anommatus* and *Cerylon*)
Corylophidae (p. 57 5)
Coccinellidae (p. 519)
Endomychidae (p. 495, g. *Endomychus* and *Lycoperdina*: p. 559, g. *Sphaerosoma* and *Symbiotes*: p. 508, g. *Mycetaea*)
Merophysiidae (p. 510), g. *Holoparamoecus*)
Lathridiidae (p. 507)
Cisidae (p. 554)
'Heteromera'
Mycetophagidae (p. 564, as Mycetophaginae)
Colydiidae (p. 508, g. *Myrmecoxenus* and *Ditoma*: p. 453, g. *Orthocerus*: p. 531, g. *Synchita* and *Cicones*: p. 501, g. *Endophloeus*: p. 516, g. *Langelandia* and *Colydium*: p. 517, g. *Aulonium*, *Tereduis* and *Oxylaemus*)
Tenebrionidae (p. 311 and p. 301, g. *Lagria*)
Tetratomidae (p. 319, g. *Tetratoma*)
Salpingidae (p. 303, as Pythidae)
Pythidae (p. 304, g. *Pytho*)
Pyrochroidae (p. 301)
Melandryidae (p. 317)
Scraptiidae (p. 308, g. *Scraptia* and *Anaspis*)
Mordellidae (p. 308)
Rhipiphoridae (p. 307)
Oedemeridae (p. 302)
Meloidae (304)
Anthicidae (p. 306)
Aderidae (p. 306, as *Hylophilus*)
Chrysomeloidea
Cerambycidae (p. 374, whole of Longicornia)
Bruchidae (p. 387, as Lariidae)
Chrysomelidae (p. 388)
Curculionoidea
Nemonychidae (p. 161, g. *Rhinomacer*)
Anthribidae (p. 240)
Attelabidae (p. 161, g. *Attelabus*, *Apoderus* and *Rynchites*)
Apionidae (p. 163)
Curculionidae (p. 158)
Scolytidae (p. 231)

STREPSIPTERA
Stylopidae
Halictophagidae
Elenchidae

How to use the key

The key is sub-divided into several individual sections (Main key (A), Keys B-K for Coleoptera and Key L for Strepsiptera). When attempting identification, the user should begin at Key A and proceed, through other keys as indicated in the appropriate choice within couplets. For other than the largest beetles, a binocular microscope will be needed.

Key A. Main key

A1. Elytra (wing cases) covering most of the abdomen, meeting on the centre-line, leaving at most 1-2 abdominal segments exposed A4

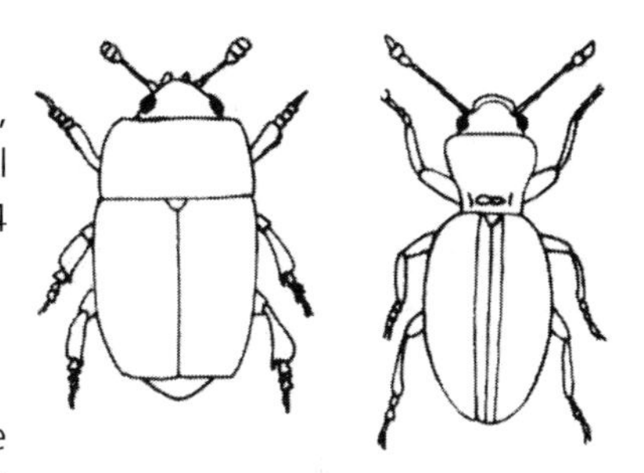

\- Elytra shortened, absent or not meeting on the centre-line A2

A2. Elytra reduced to knob-like structures; hind wings fully developed .... Order Strepsiptera ... Key L (page 41)

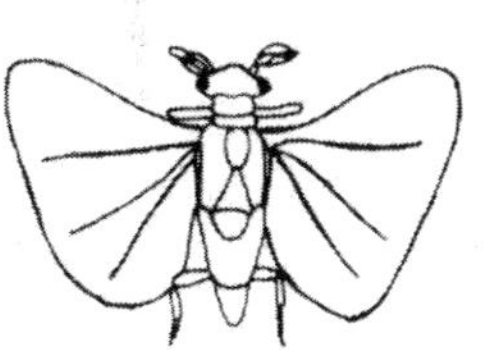

\- Beetles without these very distinctive characters A3

A3. Elytra truncate, exposing 3-6 abdominal segments, but always meeting on the centre-line .... Key C (page 12)

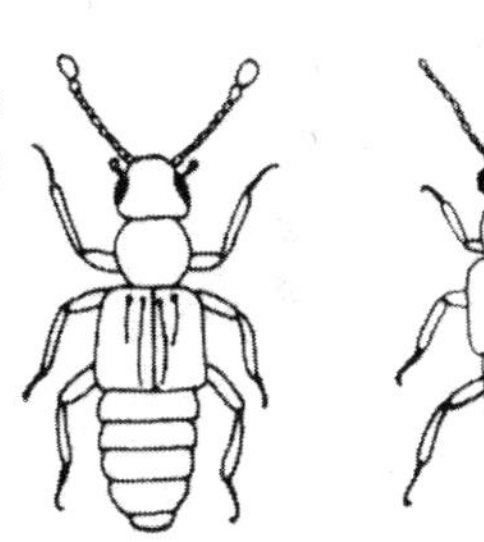

\- Elytra completely absent, or reduced and not meeting on the centre-line. Never truncate, as above .... Key B (page 10)

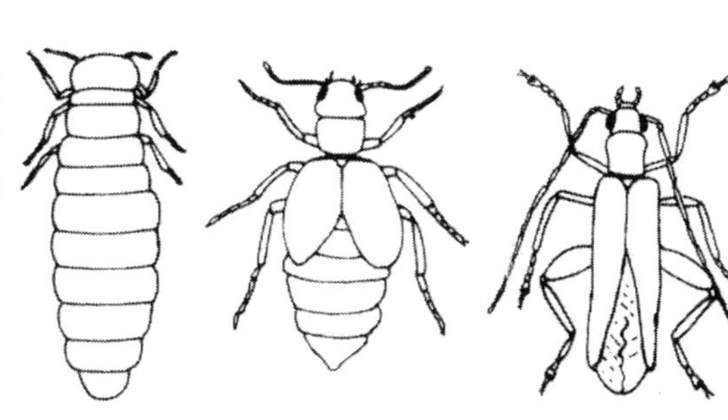

A4. Middle and hind legs shorter than front legs, flattened. Oval beetles with 4 eyes .... Key F (page 16)

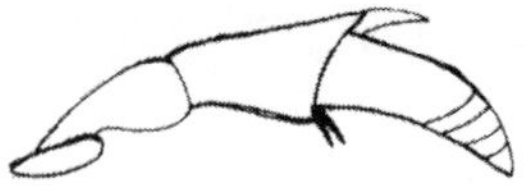

- Legs not as above A5

A5. Hind tarsi with 5 segments, evenly tapering, and bearing fine pale swimming hairs (it is often easier to see this character if a drop of water is placed on the tarsus). Oval, boat-shaped beetles ...... Key F (page 16)

- Hind tarsi with fewer than 5 segments, **or** not evenly tapering, **or** without swimming hairs A6

A6. Antennae of unusual shape (all such antennae are illustrated here) .... Key E (page 14)

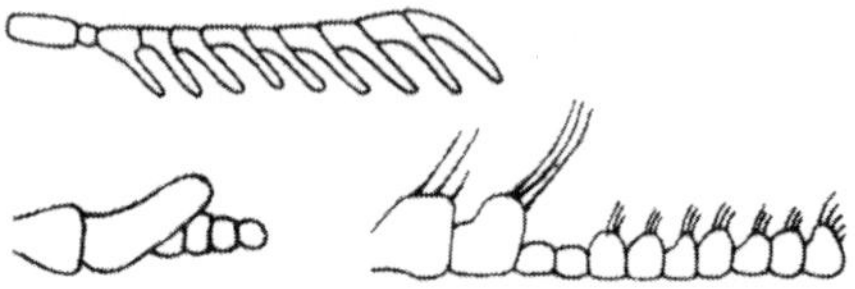

- Antennae of form other than those illustrated above A7

A7. Antennae with a club composed of plates (lamellate club). Such antennae are often strongly asymmetrical .... Key D (page 13)

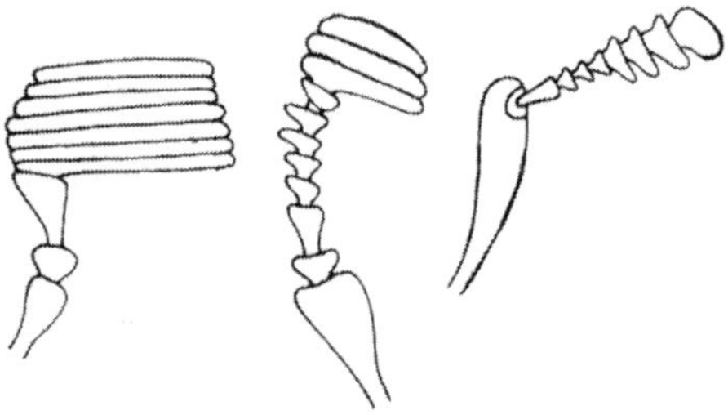

- Antennae thread-like, or with a club not composed of plates. Usually symmetrical A8

Beware: beetles have antennae with more than 3 segments. If it appears to have only 3 segments, you are probably looking at the palps.

A8. Antennae thread-like (filiform), **or** serrated, **or** gradually expanded, but never with a sudden increase in width in the apical segments (clubbed) A9

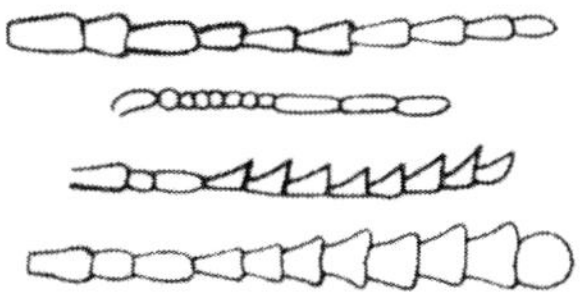

\- Antennae with a club (a sudden increase in width in the apical segments). Usually symmetrical A10

Note: Marginal cases are keyed out in both groups.

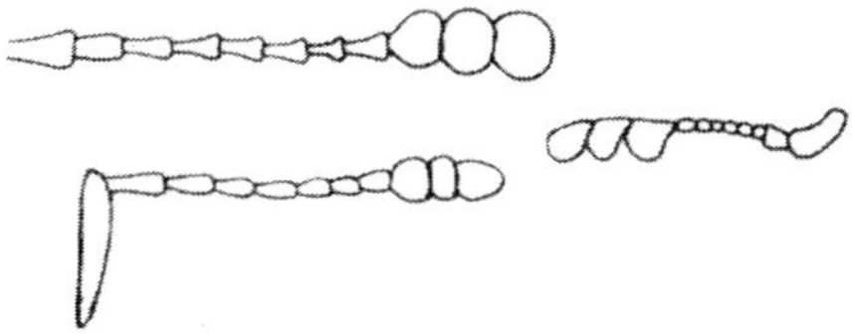

A9. Hind tarsi with some segments lobed or bilobed. See note below .... Key G (page 18)

\- Hind tarsi with all segments simple, without lobes .... Key H (page 22)

Note: Marginal cases are keyed out both ways.

A10. Hind tarsi with some segments lobed or bilobed. See note below .... Key J (page 28)

\- Hind tarsi with all segments simple, without lobes .... Key K (page 32)

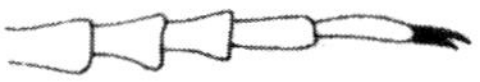

Note: Marginal cases are keyed out both ways.

Note: Tarsal characters are often easier to see, particularly in small beetles, if a drop of water is placed on the tarsus.

Simple tarsi may have conical segments, but they have radial symmetry.

Bilobed tarsi have segments developed on the underside only.

Key B

Beetles with elytra absent or greatly reduced, not meeting on centre-line of abdomen, and rounded or pointed apically.

B1. Elytra completely absent B2

- Elytra present, but reduced B5

B2. Antennae with 11 segments, thread-like. Tarsi with 5 segments, some bilobed B3

- Other types of antennae and tarsi .... **Beetle larvae**

B3. Head covered by thorax (shown arrowed). Length of antennae twice distance between eyes. Female glow-worms .... **LAMPYRIDAE**

Thorax

- Head not covered by thorax B4

B4. Antennae short, about equal to distance between eyes. Beetle 12-15 mm long .... Female **DRILIDAE**

Thorax

- Antennae or size different .... **Beetle larvae**

B5. Antennae pectinate (comb-like) .... **RHIPIPHORIDAE**

- Antennae filiform (thread-like) B6

B6. Tarsi with all segments simple, not lobed or bilobed B7

- Tarsi with some segments lobed B8

Note: Tarsal characters are often easier to see, particularly in small beetles, if a drop of water is placed on the tarsus.

Simple tarsi may have conical segments, but they have radial symmetry.

Bilobed tarsi have segments developed on the underside only.

B7. Front and middle tarsi with five segments, hind with 4 segments. Broad beetles; elytra well separated at apex .... **MELOIDAE**

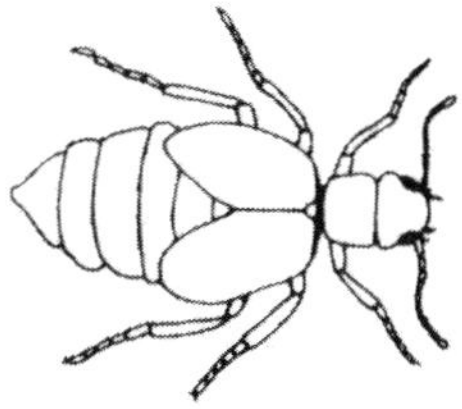

- All tarsi with 5 segments. Elongate beetles; elytra narrowly separated at apex .... **LYMEXYLIDAE**

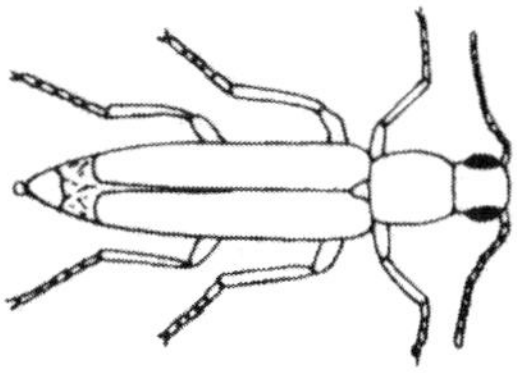

B8. Head covered by thorax. Male glow-worm, *Phosphaenus* .... **LAMPYRIDAE**

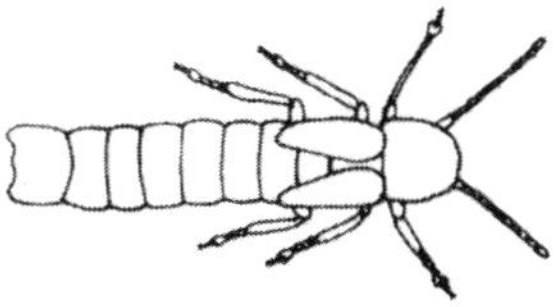

- Head not covered by thorax B9

B9. Apparent number of tarsal segments.

Note: In this key, small cylindrical tarsal segments (arrowed) immediately following a bilobed segment are ignored. Claws do not count as segments.

Legs:	front	middle	hind	
-	4	4	4	.... **CERAMBYCIDAE**
-	5	5	4	.... **OEDEMERIDAE**
-	5	5	5	.... **CANTHARIDAE**

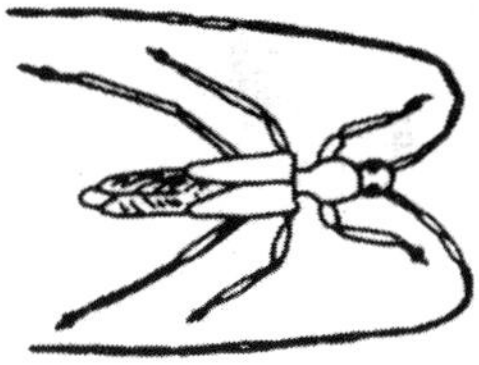

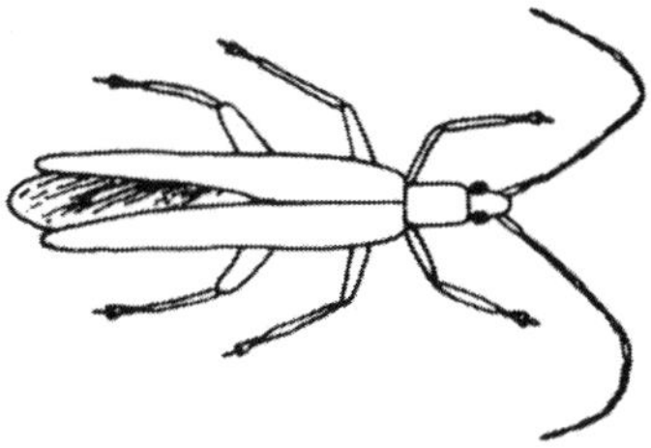

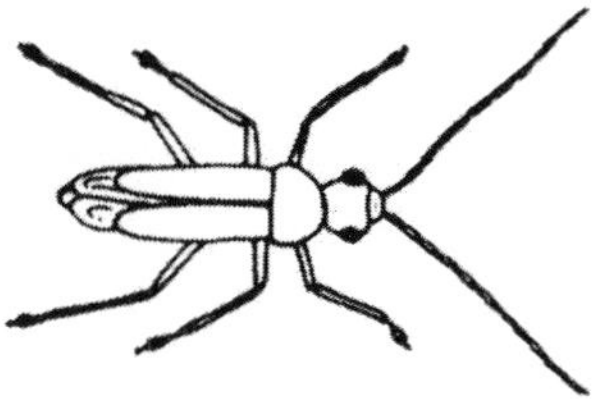

Key C

Beetles with elytra truncate, exposing 3-6 segments of abdomen. (Care should be taken with specimens taken in water traps, which may have their abdomens distended.)

C1. Minute beetles, less than 1.4 mm long, excluding antennae. Antennae with first 2 segments stout, segments 3 onwards very thin, and a loose 3-segmented club, often with hairs .... **PTILIIDAE**

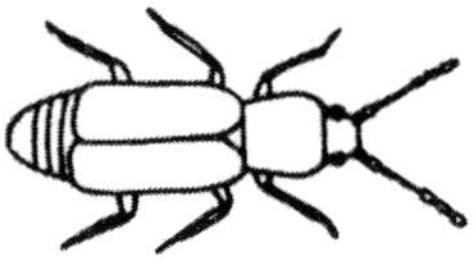

- Larger beetles, or antennae different C2

C2. Antennae short, thick, with only 5 segments .... **PSELAPHIDAE**

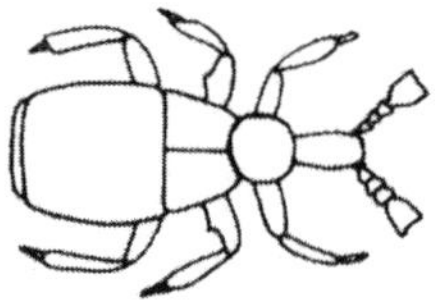

- Antennae with more (usually 9-11) than 5 segments C3

C3. Elytra short, shorter than the exposed part of the abdomen C7

- Elytra at least as long as the exposed part of the abdomen C4

C4. Beetles over 10 mm or under 1.4 mm (excluding antennae) A4

- Beetles between 1.4 and 10 mm C5

Note: Marginal cases can go either way.

C5. Hind tarsi with strongly bilobed segments, or tibiae expanded and serrated or toothed on the outside A4

Note: If ocelli are present (C6 below) go to C7.

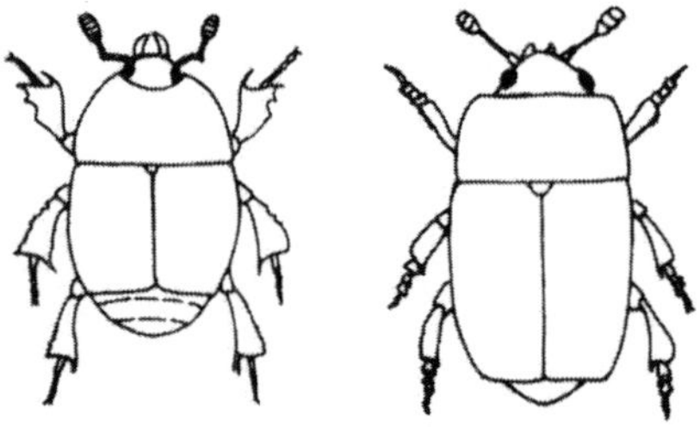

- Hind tarsi without any strongly bilobed segments; tibiae slender C6

C6. Ocelli present (ocelli are lens-like simple eyes on the top of the head) C7

Note: Ocelli may be very small, and sometimes only one is present.

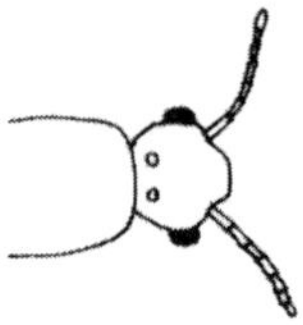

- Ocelli absent A4

C7. Antennae and palps clubbed .... **PSELAPHIDAE**

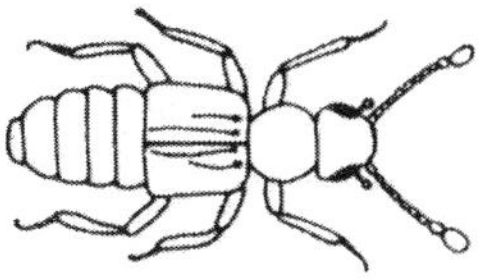

- Antennae may be weakly clubbed, but never **both** antennae and palps .... **STAPHYLINIDAE**

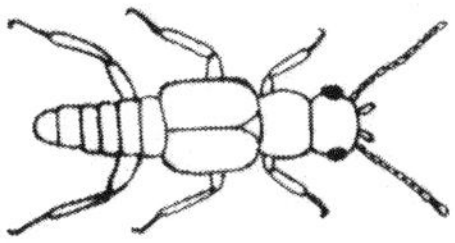

Key D

Antennae with an asymmetric, lamellate club.

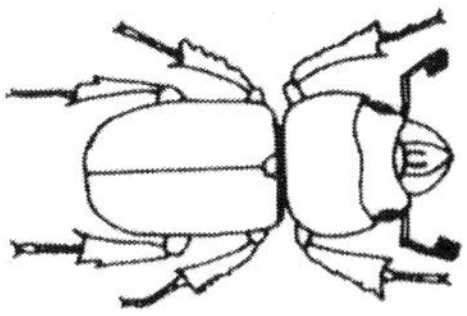

D1. Antennae very obviously 'elbowed', with a very long scape (extended basal segment) clearly visible from above. Males of some species with huge antler-like mandibles. Stag beetles .... **LUCANIDAE**

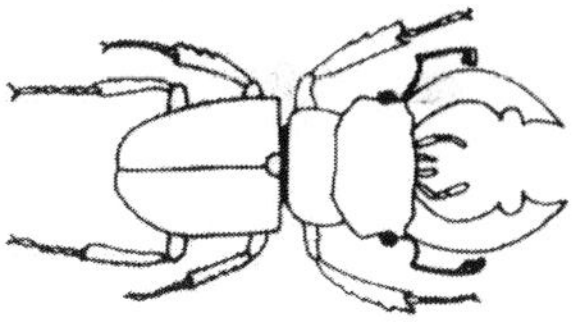

Note: Make sure that what is taken for a scape really consists of only one segment.

- Scape much shorter than this, generally not visible from above D2

Note: The lucanid *Sinodendron* spp. will key out here. It has a thorax almost as long as broad with a concave front margin. 10-15 mm. Male with a forward-projecting horn on the head.

D2. Elytra short, exposing 3-4 segments of the abdomen ...... **SILPHIDAE**

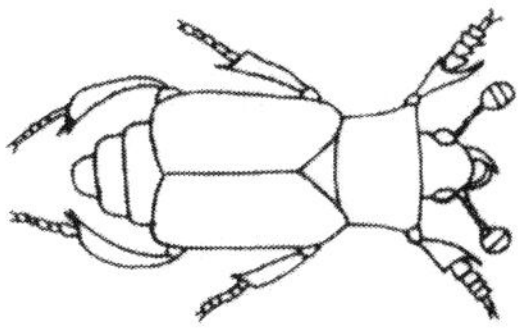

- Elytra covering all or nearly all of the abdomen D3

D3. Thorax and elytra dull, roughly sculptured .... **TROGIDAE**

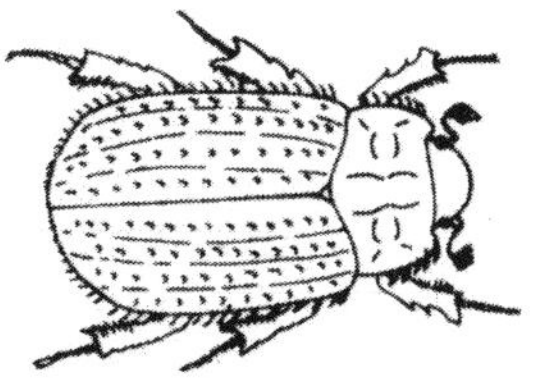

- Thorax and elytra always somewhat shining, not roughly sculptured in this way D4

D4. Antennae with 11 segments. Thorax produced forward into points **or** (a) head with rear-pointing horns **or** (b) mandibles very large, clearly visible from above .... **GEOTRUPIDAE**

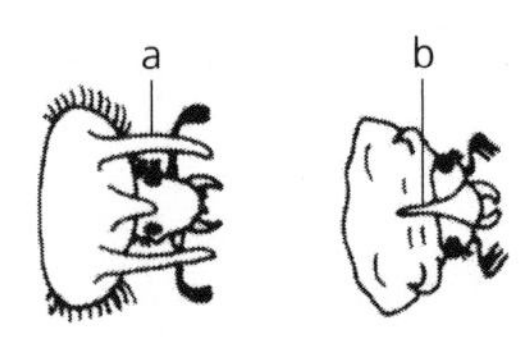

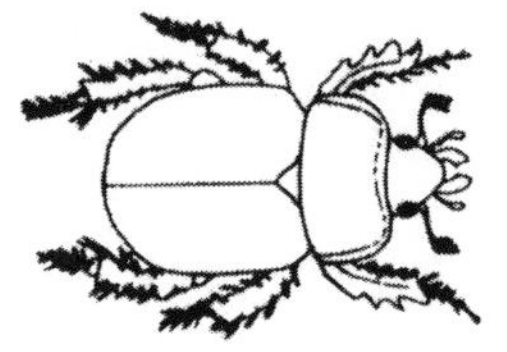

- Antennae with 9 or 10 segments, without the features shown above .... **SCARABAEIDAE**

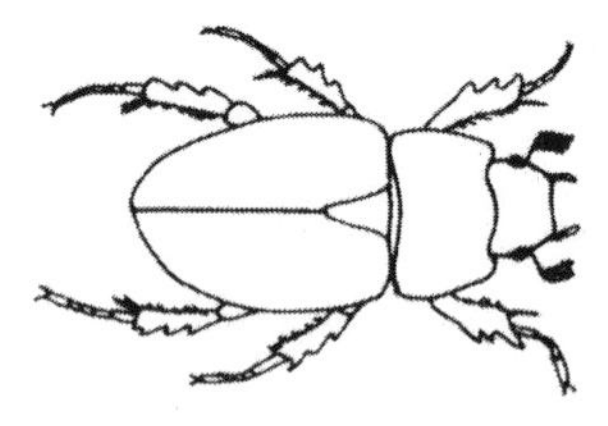

Key E

Beetles with elytra covering most of the abdomen. Antennae of unusual shape.

E1. Antennae pectinate E3

- Antennae of other shapes E2

E2. Antennae as illustrated. Tarsi with 4 simple (not bilobed) segments .... **HETEROCERIDAE**

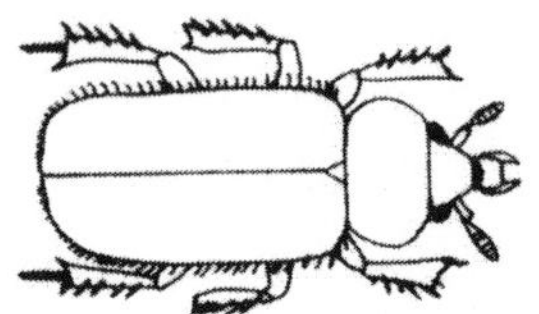

- Antennae as illustrated. Tarsi with 5 simple (not bilobed) segments .... **DRYOPIDAE**

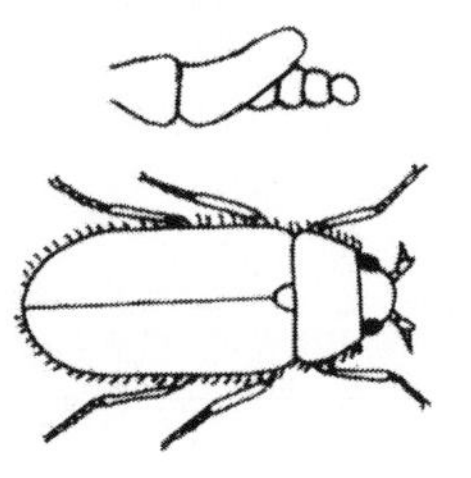

E3. All tarsi with 5 segments (check hind tarsi carefully) E4

- Front and middle tarsi with 5, hind tarsi with 4 segments .... **PYROCHROIDAE**

E4. Tarsi with 5 segments, some bilobed E5

- Tarsi with simple segments E6

Note: Tarsal characters are often easier to see, particularly in small beetles, if a drop of water is placed on the tarsus.

Simple tarsi may have conical segments, but they have radial symmetry.

Bilobed tarsi have segments developed on the underside only.

E5. Eyes quite small, approximately circular .... **DRILIDAE**

- Eyes very large, much higher than wide. Antennae inserted very low (see G2) .... **BUPRESTIDAE**

E6. Prosternum (underside of 1st segment of thorax) with a pointed rearward extension, fitting into a cavity in 2nd segment – see Fig. 3, p. 3 (and H14) .... **ELATERIDAE** and **EUCNEMIDAE**

- Prosternum without this extension. Thorax hood-shaped (see H14) .... **ANOBIIDAE**

Key F

Aquatic beetles.

F1. Hind and middle tarsi greatly flattened, half the length of the front tarsi. Beetles with 4 eyes and short, unusual antennae .... **GYRINIDAE**

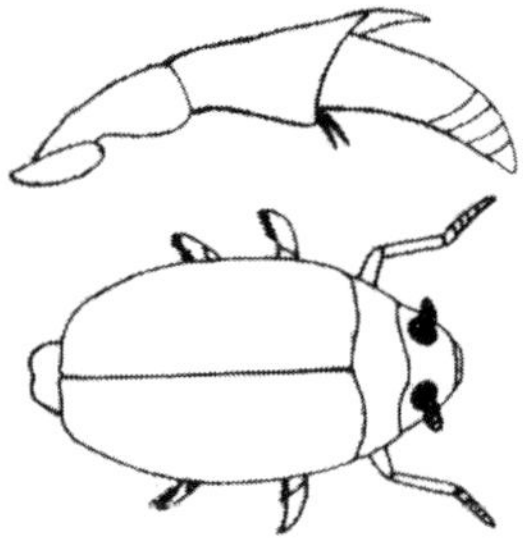

\- Tarsi not as above; beetles with 2 eyes F2

F2. Palps long, at least 2/3 the length of the antennae, which are clubbed F3

\- Palps shorter than 2/3 the length of the antennae F5

Note: Palps have three segments, antennae have many more (usually 11).

F3. Thorax with widest part close to base .... **HYDROPHILIDAE**

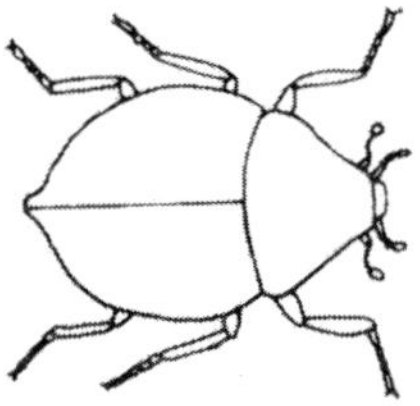

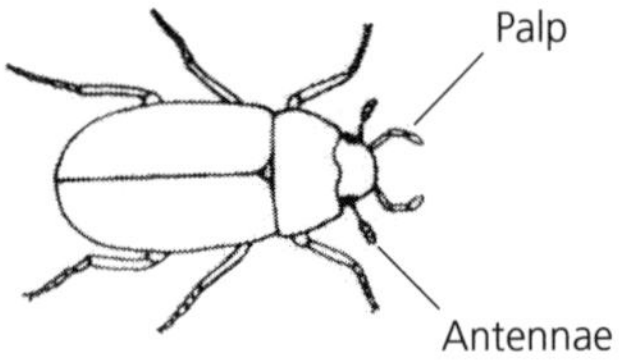

Palp

Antennae

\- Thorax with widest part about the middle F4

F4. Club of antenna (which is hairy, in contrast to the rest of the antenna) with 3-4 segments; abdomen with 4-5 visible segments (count from between hind legs to the apex, ignoring genitalia, which are soft parts, light in colour, and usually retracted); thorax often with several evenly spaced longitudinal furrows .... **HYDROPHILIDAE**

\- Antennal club (which is hairy, in contrast to rest of antenna) with 5 segments: abdomen with 6-7 segments (counted as above); thorax with indentations, or a median furrow but never with several evenly spaced furrows .... **HYDRAENIDAE**

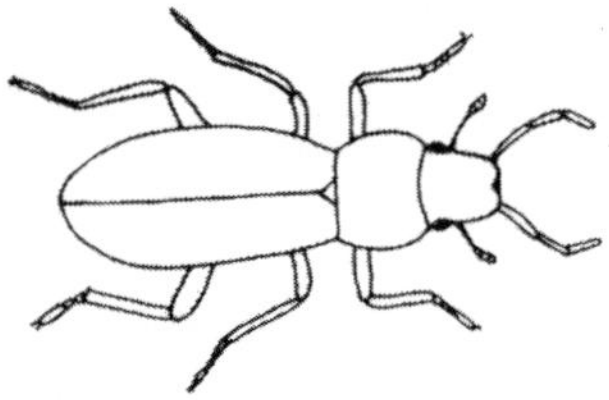

F5. Hind coxae plate-like, covering half the underside of the abdomen .... **HALIPLIDAE**

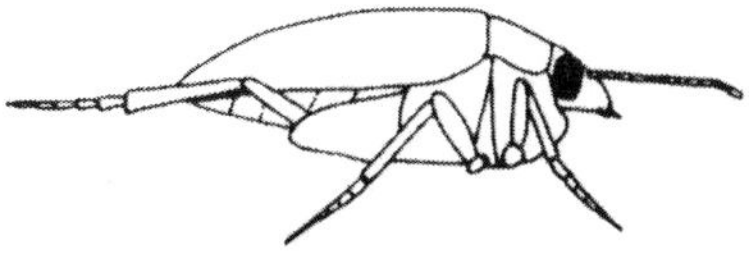

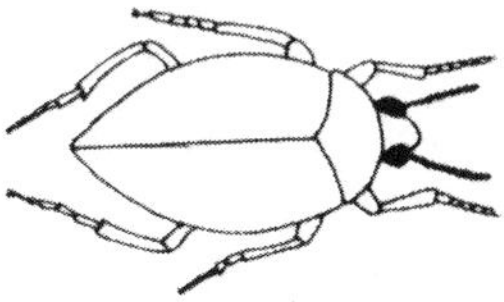

- Hind coxae much smaller than this, most of the hind femora exposed F6

F6. Beetles over 8 mm long (excluding antennae) F7

- Beetles up to 8 mm long F8

F7. Underside of beetle very convex .... **HYGROBIIDAE**

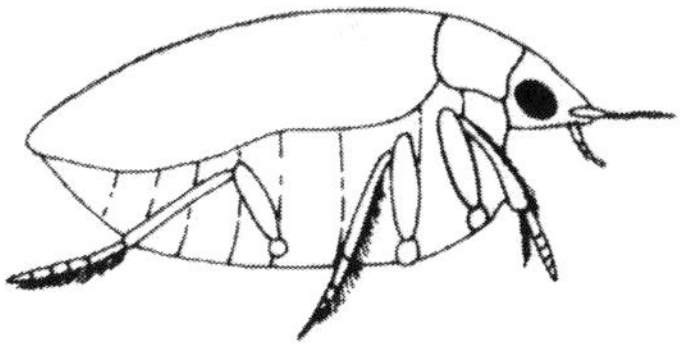

- Underside of beetle much flatter .... **DYTISCIDAE**

F8. Middle antennal segments cup-shaped, some as wide or wider than long at the front margin and with lobes in the male F9

- All antennal segments more cylindrical, longer than wide at front margin .... **DYTISCIDAE**

F9. Beetles 3.5-5 mm long (excluding antennae) .... **NOTERIDAE**

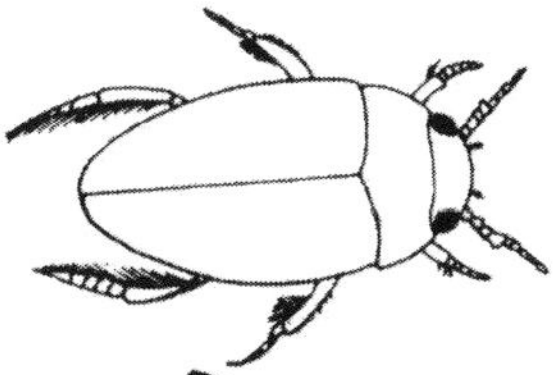

- Beetles less than 3 mm long .... **DYTISCIDAE**

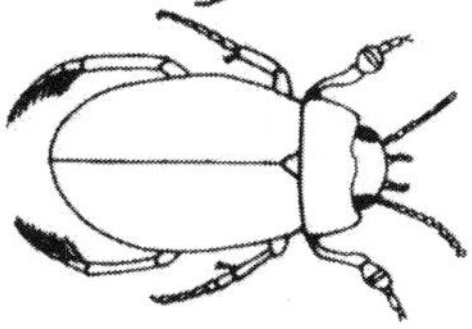

Key G

Beetles with elytra covering most of the abdomen. Antennae filiform, or serrate, and hind tarsi with some segments lobed or bilobed.

G1. Apparent number of tarsal segments; five combinations are listed below G2

Note: In this key, small cylindrical tarsal segments (arrowed) immediately following a bilobed segment are ignored. Claws do not count as segments.

Note: Tarsal characters are often easier to see, particularly in small beetles, if a drop of water is placed on the tarsus.

Legs:	front	middle	hind	
	5	5	5	.... G2
	5	4	4	.... G10
	5	4	4	.... G16
	4	4	3	.... **ADERIDAE**
	3	3	3	.... **ENDOMYCHIDAE**

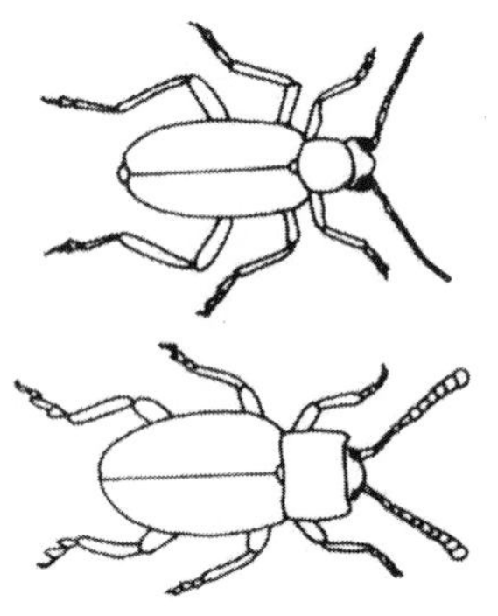

G2. Eyes oval, over twice as high as wide, occupying most of head. Antennae often serrated .... **BUPRESTIDAE**

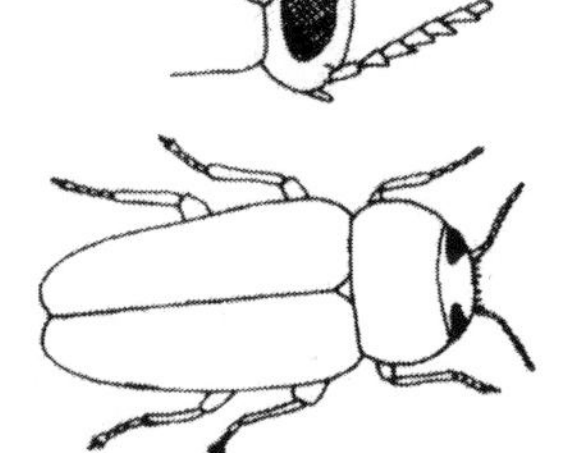

- Eyes approximately circular, or absent G3

G3. Eyes completely absent. Brownish-orange beetles .... **LEPTINIDAE**

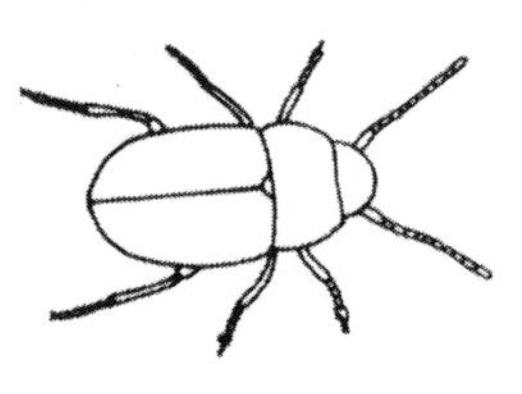

- Eyes present G4

G4. Antennae very short; head covered by thorax .... **LAMPYRIDAE**

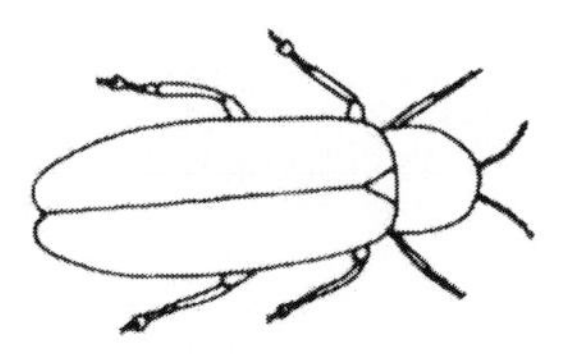

- Antennae much longer; head not completely covered by thorax G5

G5. Thorax sculptured in squarish patterns: elytra with obvious longitudinal ridges .... **LYCIDAE**

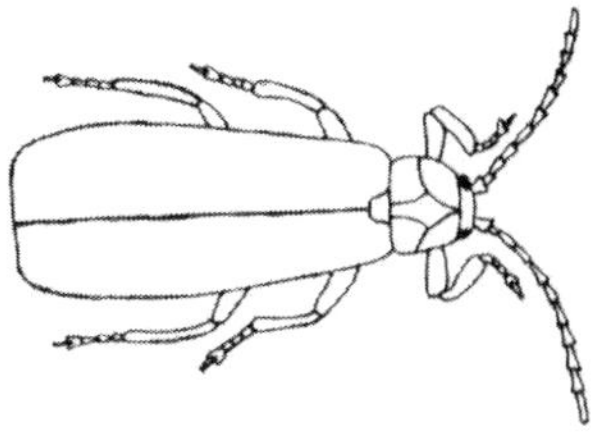

- Thorax not sculptured in this way, usually smooth G6

G6. Antennae expanded; thorax with outstanding hairs .... **CLERIDAE**

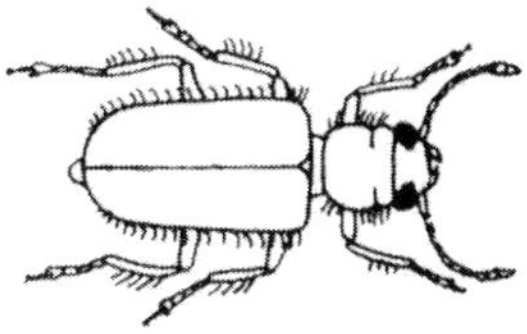

Note: Some Melyridae, H10, may key out here if hind tarsi appear to be bilobed. They do not have antennae expanded.

- Antennae not serrated; thorax with only short, flattened hairs, or bare. Front tibiae without a distinct notch on inner side G7

Note: If a notch is present, try Carabidae, H3.

G7. Beetles over 8 mm long (excluding antennae) G8

- Beetles up to 8 mm long G9

G8. Orange-brown beetles, completely covered with dense flattened light yellow hairs. Beetle completely dull .... **DASCILLIDAE**

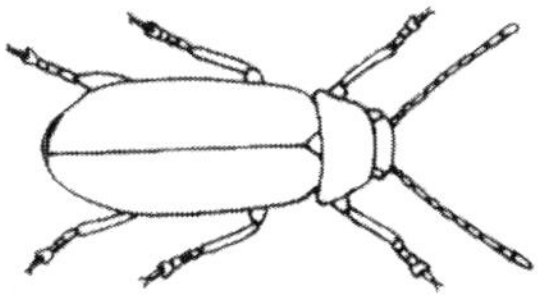

- Yellowish-brown beetles, often with black markings, lightly covered with pale yellow hairs, with the ground colour clearly visible everywhere. Always somewhat shining .... **CANTHARIDAE**

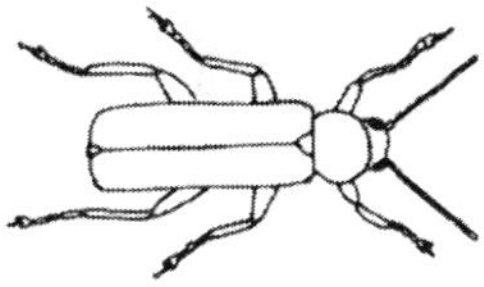

G9. Hind femora as long as the width across both elytra. Usually over 4 mm long (excluding antennae) .... **CANTHARIDAE** (see G8)

Note: The carabid *Demetrias* spp. will key out here. Head black; thorax reddish-yellow; elytra yellow often with dark markings. 'Waist' narrow (rear of thorax is much narrower than the elytra).

- Hind femora just half as long as the width across both elytra. Beetles 1.5-4 mm long .... **SCIRTIDAE**

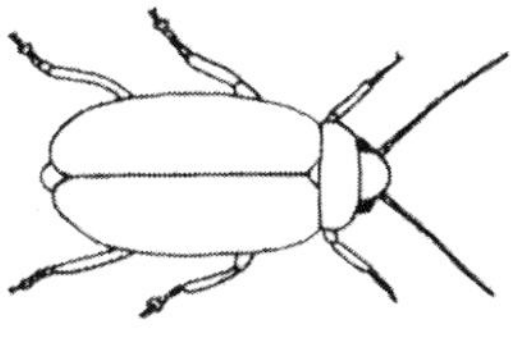

Note: Some Buprestidae, G2, may key out here; they have front of head obviously cleft or concave. Anobiidae may also key out here; they have a hood shaped thorax, see H14.

G10. Antennae serrate .... **PYROCHROIDAE**
(see E3)

- Antennae filiform G11

G11. Eyes deeply notched in front to accommodate antennae G12

- Eyes approximately circular, or at most very slightly indented G13

G12. Beetles at least 6 mm long (excluding antennae) .... **MELANDRYIDAE**
(see G15 below)

- Beetles up to 4 mm long .... **SCRAPTIIDAE**

Note: Some Tenebrionidae may key out here; they have a black thorax and yellow elytra, both with outstanding hairs (formerly Lagriidae).

G13. Thorax extended forwards over the head in the form of a horn .... **ANTHICIDAE**

- Thorax without a horn G14

G14. Back of head not shaped to fit front of thorax, leaving a distinct 'neck' .... **ANTHICIDAE**

- Back of head shaped to fit thorax; beetles without a distinct 'neck' G15

G15. Thorax longer than broad. Elytra with 3 distinct longitudinal ridges. (Elytra not fully covering abdomen, or not meeting at centre-line.) .... **OEDEMERIDAE**

- Thorax shorter than broad. Elytra smooth, or with about 10 very indistinct lines .... **MELANDRYIDAE**

G16. Head extended in front of eyes as a rostrum (beak) longer than width of head .... **NEMONYCHIDAE**

Note: Some weevils (Curculionidae) may key out here, but these have clubbed antennae with an extended basal segment.

- Head not extended in this way G17

G17. Antennae inserted into deep pits. Brownish beetles with raised bumps on thorax and abdomen; densely covered with brownish flattened hairs, with patches of whitish hairs .... **ANTHRIBIDAE**

- Antennae not inserted into deep pits G18

G18. Eyes notched in front to accommodate antennae, or eyes partly surrounding antennae G19

- Eyes not shaped in this way G20

G19. Hind tibiae about half as long as the width across both the elytra. 2-2.5 mm long (excluding antennae) .... **BRUCHIDAE**

Note: If under 2 mm, try Aderidae, G1.

- Hind tibiae well over half the width across both elytra .... **CERAMBYCIDAE**

G20. 2nd antennal segment very short, about 1/3 length of 3rd segment .... **CERAMBYCIDAE**

- 2nd antennal segment at least 1/2 as long as 3rd segment .... **CHRYSOMELIDAE**

Key H

Beetles with elytra covering most of the abdomen. Antennae filiform, expanded or serrate, and hind tarsi without any lobed or bilobed segments.

H1. Antennae very long, much longer than the front legs, with the basal segment somewhat swollen and as long as or longer than the next two segments together; thorax with the front corners protruding and pointed .... **CUCUJIDAE**

\- Antennae without these characteristics H2

H2. Number of tarsal segments.

Note: Tarsal characters are often easier to see if a drop of water is placed on the tarsus.

Note: Claws do not count as segments.

Legs: front	middle	hind	
5	5	5	.... H3
5	5	4	.... H17
4	4	4	.... H22
4	4	3	.... **ADERIDAE** (see G1)
3	4	4	.... H26
3	3	3	.... **LATHRIDIIDAE** (see K49)

H3. Hind trochanters large, strongly projecting, always longer than the diameter of the femur, clearly visible from behind the beetle even when card mounted .... **CARABIDAE**

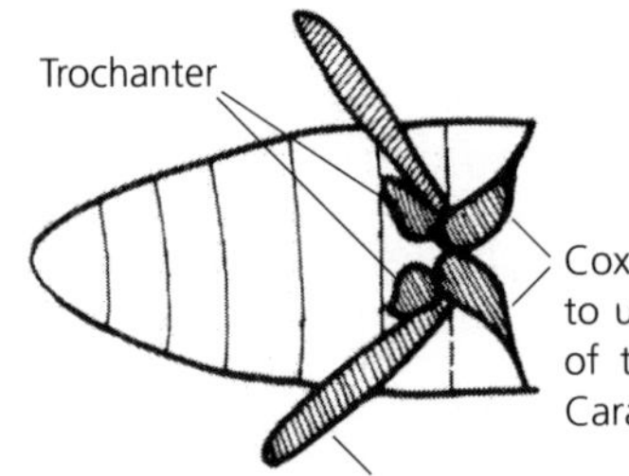

\- Hind trochanters small, not longer than the diameter of the femur, and not projecting H4

H4. End of the abdomen pointed, extending well beyond end of elytra (take care with water trapped beetles which may be distended) H5

\- End of the abdomen rounded, usually covered by elytra H6

H5. Oval beetles .... **SCAPHIDIIDAE**

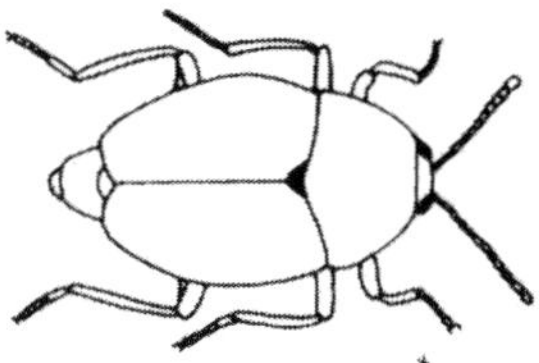

- Very elongate beetles .... **LYMEXYLIDAE**

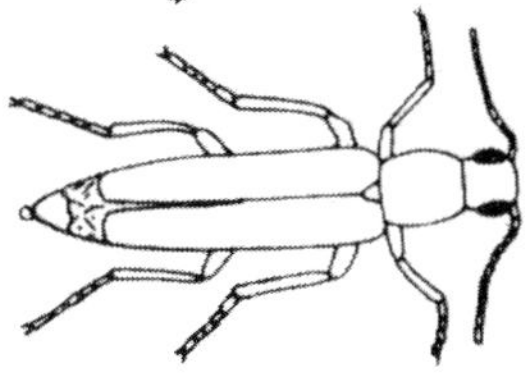

  Note: 1 species of Scydmaenidae, K23, which is much less elongate, may key out here. It has clear indentations on hind margin of thorax.

H6. Front tarsi with segments strongly bilobed H7

  Note: Tarsal characters are often easier to see, particularly in small beetles, if a drop of water is placed on the tarsus.

  Simple tarsi may have conical segments, but they have radial symmetry.

  Bilobed tarsi have segments developed on the underside only.

- Front tarsi with segments all simple H8

H7. Oval beetles about 4 mm long (excluding antennae) with slender front tibiae .... **EUCINETIDAE**

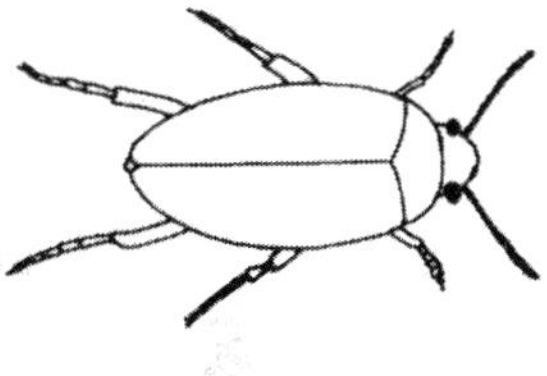

  Note: One species of Melyridae, H15, may key out here; it has a single lobe on the second segment of the front tarsi.

- Beetles over 9 mm long, with front tibiae expanded and with a sharp point .... **SILPHIDAE**

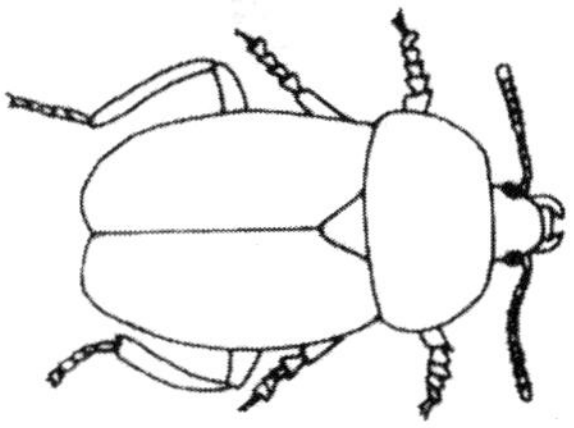

H8. Very elongate beetles (4 times longer than broad) with the rear ends of the elytra somewhat pointed: covered with flattened pubescence (hairs) .... **MELYRIDAE**

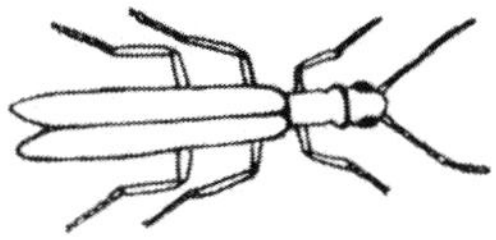

- Beetles not so elongate, and ends of elytra not shaped in this way H9

H9. Very short, fat beetles; thorax short, about 1/3 as long as broad .... **PSEPHENIDAE**

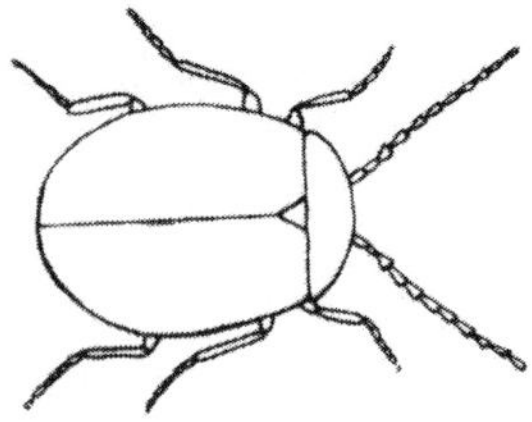

- Beetles more elongate than this, or thorax more elongate H10

H10. Beetles with outstanding long hairs on thorax. Hind corners of thorax rounded .... **MELYRIDAE**

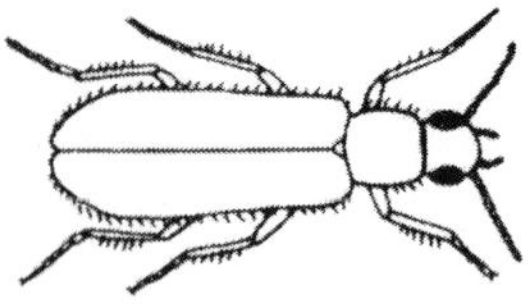

- Beetles without long hairs on thorax (some tiny, very rounded beetles have a few scattered erect hairs) H11

Head

Prosternum

H11. Prosternum (underside of 1st segment of thorax) with a pointed rearward extension that fits into a cavity in the 2nd segment. Hind corners of thorax sharp .... **ELATERIDAE** and **EUCNEMIDAE**

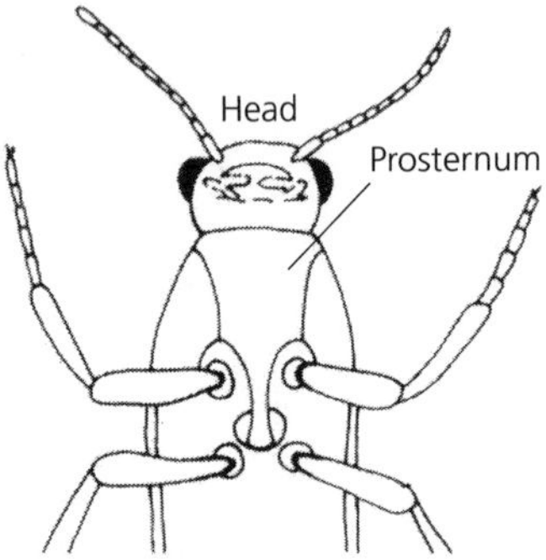

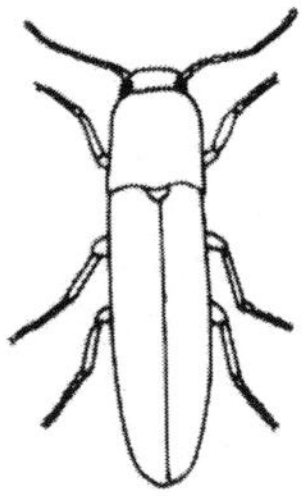

- Prosternum without this extension. Hind corners of thorax usually rounded H12

H12. Maximum width of the thorax about the same as the elytra H13

- Thorax much narrower than elytra H16

H13. Antennae distinctly shorter than the front legs .... **ELMIDAE**

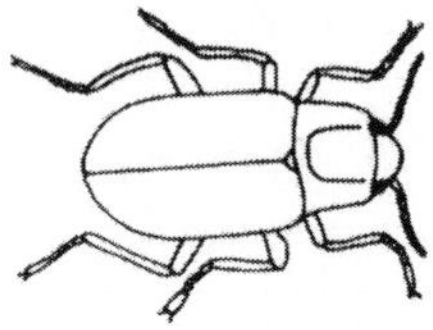

- Antennae about the same length as front legs H14

H14. Thorax shaped like a hood, covering the head; antennae often with the apical segments lengthened. Underside of beetle with a groove to accommodate the hind femora when folded (hind coxae excavated behind .... **ANOBIIDAE**

- Thorax not hood-shaped; apical antennal segments not lengthened H15

H15. Head (across the eyes) about as wide as thorax .... **MELYRIDAE**

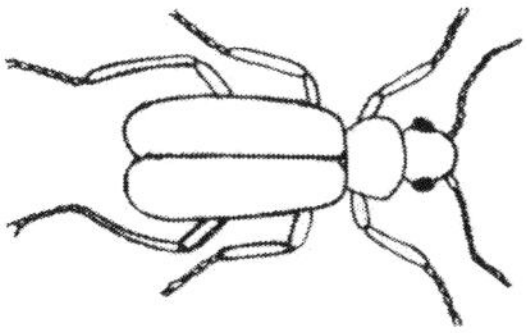

- Head (across the eyes) narrower than thorax .... **LEIODIDAE**

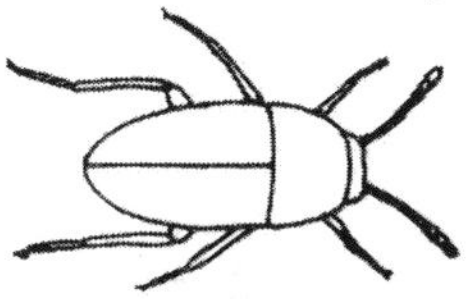

H16. Beetles with a distinct 'neck' (back of head not shaped to fit front of thorax) .... **SCYDMAENIDAE**

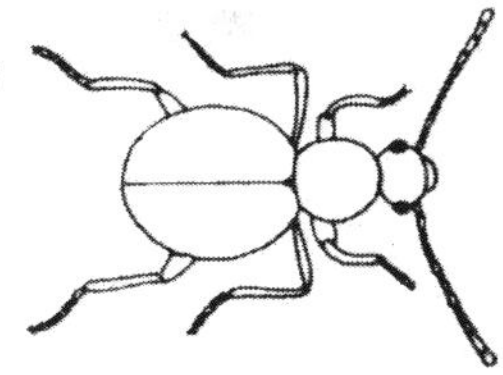

- Back of head shaped to fit front of thorax; elytra often oval .... **PTINIDAE**

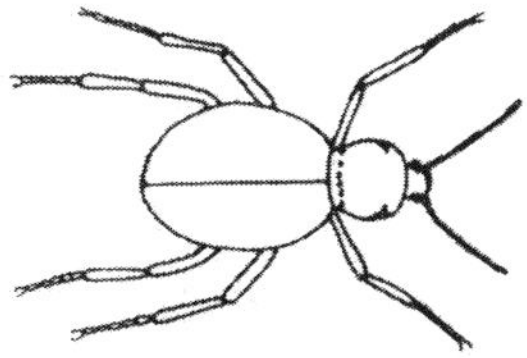

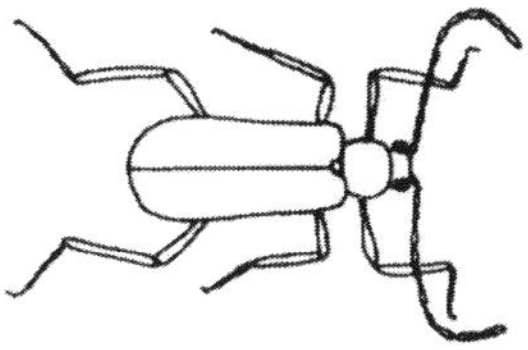

H17. Abdomen pointed, extending well beyond end of the elytra (note: abdomen pointed, not the elytra) .... **MORDELLIDAE**

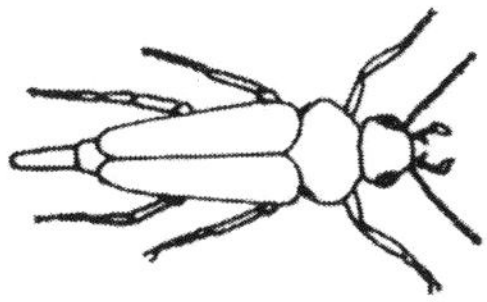

- Abdomen not pointed, and usually covered by the elytra H18

H18. Thorax at base about as wide as the front of the elytra H19

- Thorax contracted at base; and narrower (at base) than the front of the elytra H20

H19. Hind tibiae with very long spurs. Red beetles with black patches .... **MELANDRYIDAE**

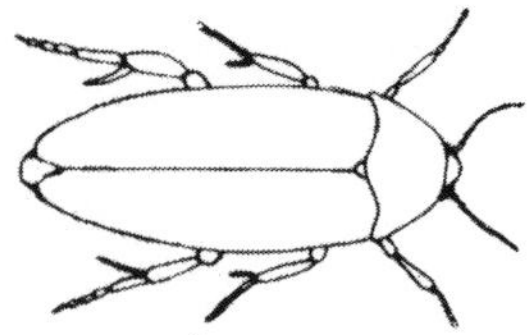

- Hind tibiae with at most only small spurs .... **TENEBRIONIDAE**

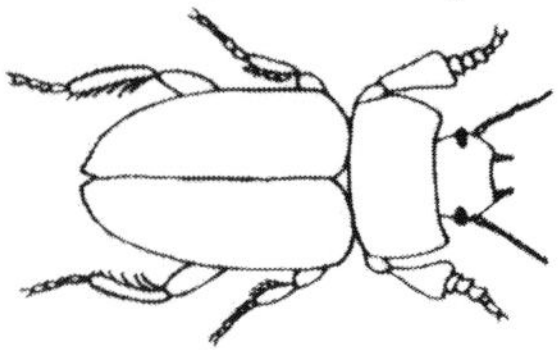

H20. Beetles at least 7 mm long (excluding antennae) H21

- Beetles less than 7 mm long .... **SALPINGIDAE**

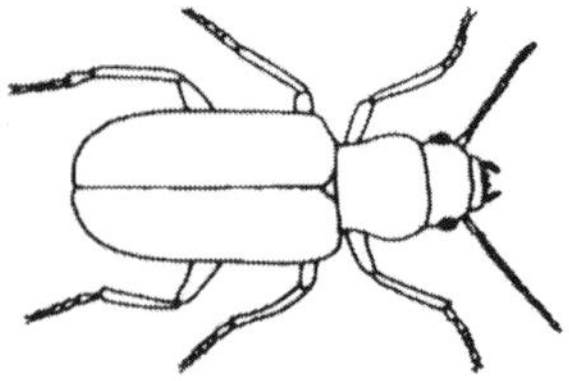

H21. Very flattened beetles with two large shallow depressions at the base of the thorax .... **PYTHIDAE**

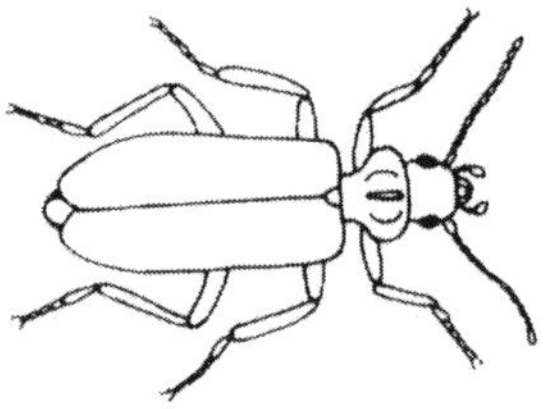

- Beetles not unusually flattened; black or green in colour .... **MELOIDAE**

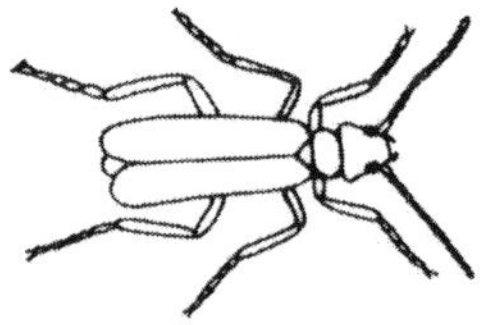

H22. Last tarsal segment of hind legs twice as long as the other segments together .... **CHRYSOMELIDAE**

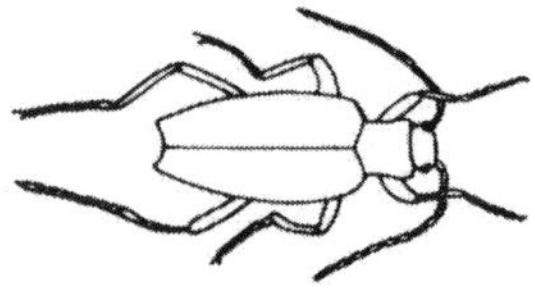

- Last tarsal segment much shorter than this H23

H23. Front tarsi with 3rd segment bilobed. 2-4.5 mm long (excluding antennae) .... **SCRAPTIIDAE**

Note: If under 2 mm, try Aderidae, H2.

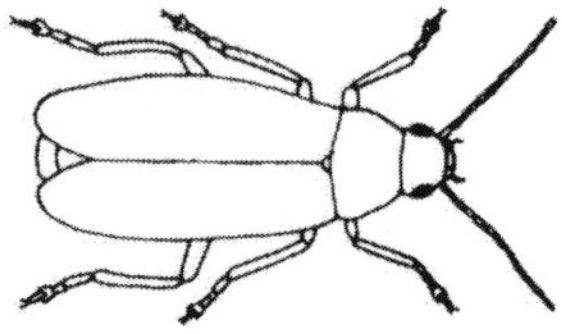

- Front tarsi with all segments simple H24

H24. Uniformly reddish-brown beetles H25

- Beetles with front of thorax yellow, or with reddish spots H26

H25. Thorax widest in front, with forward-facing points: antennae short, expanded .... **TROGOSSITIDAE**

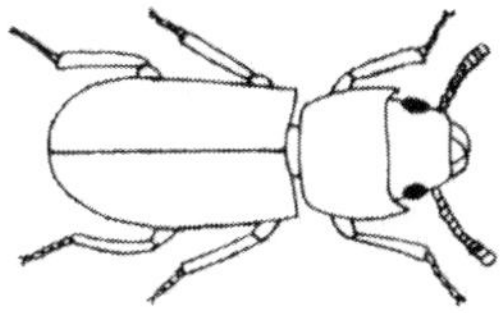

- Thorax almost rectangular; antennae long, filiform ...... **CUCUJIDAE**

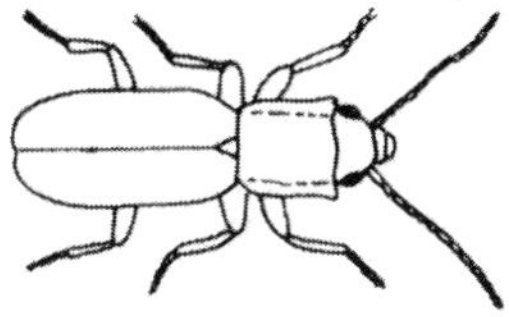

H26. Brown beetles with reddish spots on elytra .... **MYCETOPHAGIDAE**

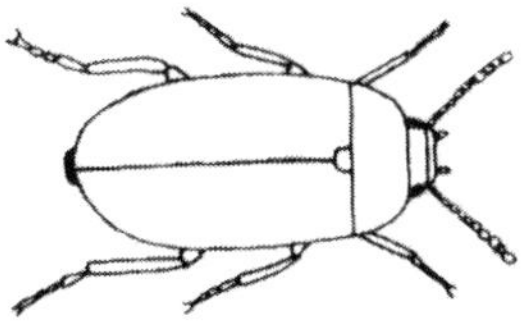

- Reddish beetles with front of thorax yellow .... **MELANDRYIDAE**

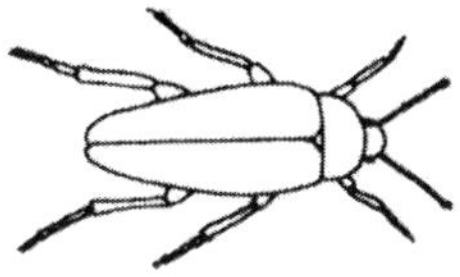

Key J

Beetles with elytra covering most of the abdomen. Antennae clubbed, and tarsi with some segments lobed beneath.

J1. Number of tarsal segments:

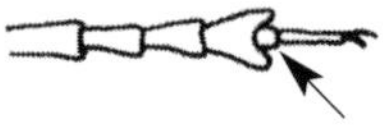

Note: In this key, small cylindrical tarsal segments (arrowed) immediately following a bilobed segment are ignored. Claws do not count as segments.

Tarsal characters may be easier to see if a drop of water is placed on the tarsus.

- Tarsi with 5 segments J2
  Tarsi with 4 segments J5
  Tarsi with 3 segments J19

J2. Thorax constricted basally, beetles with a very obvious 'waist'. Thorax with outstanding long hairs .... **CLERIDAE**

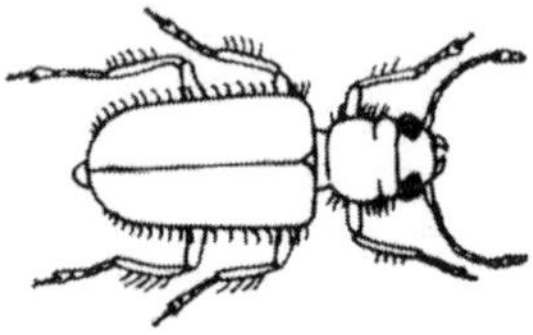

- Beetles without a distinct 'waist' between thorax and elytra J3

J3. Eyes oval, over twice as high as wide, occupying most of the height of the head .... **BUPRESTIDAE**

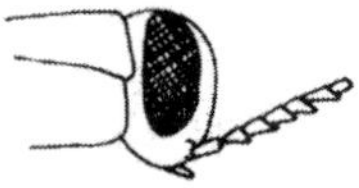

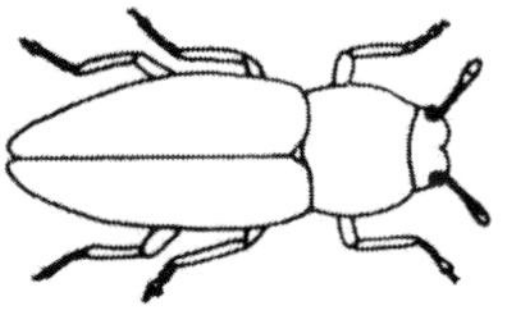

- Eyes approximately circular J4

J4. Oval beetles with a wide turned-out rim to the elytra and the thorax giving a terrapin-like appearance. Tibiae not flattened and expanded .... **PELTIDAE**

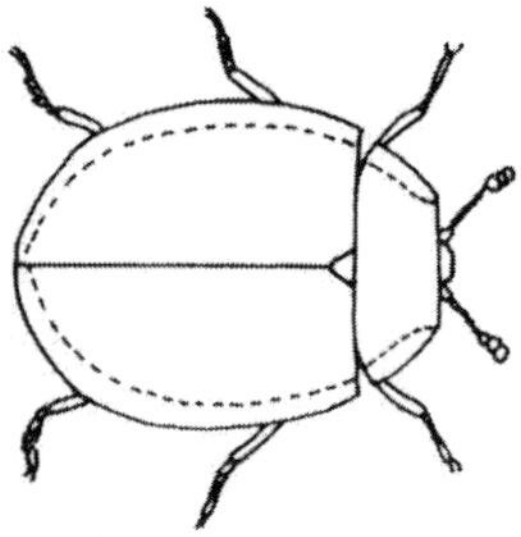

- Beetles with at most a narrow 'beading' on thorax and elytra. Tibiae flattened and expanded .... **EROTYLIDAE**

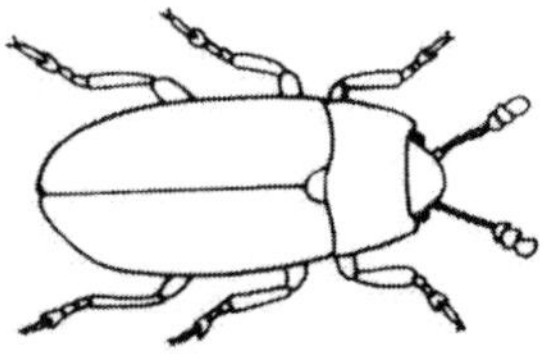

J5. Beetles with a very long 'neck'; eyes well forward of thorax .... **ATTELABIDAE**

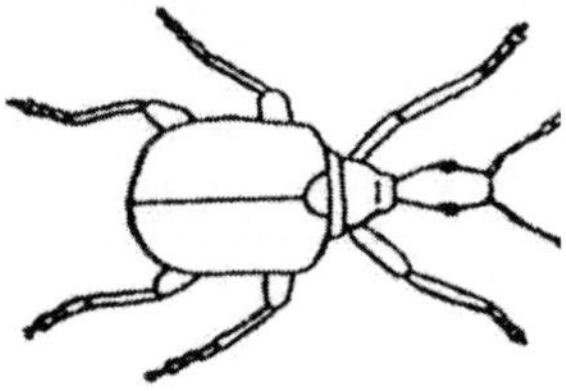

- Beetles without a long 'neck'.. J6

J6. 3rd segment of tarsi (bilobed) partly inserted into 2nd segment (also bilobed). (Antennae inserted into deep pits; eyes circular.) .... **ANTHRIBIDAE**

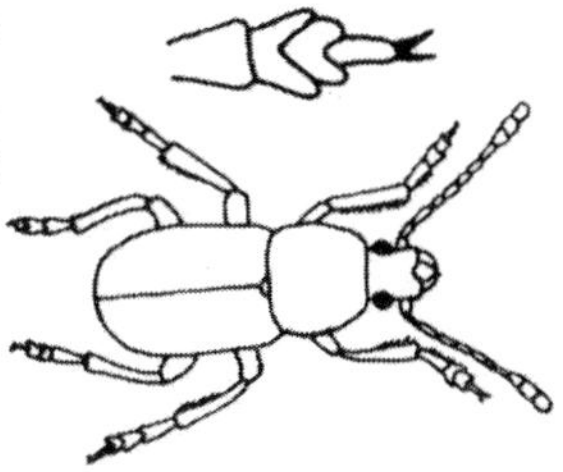

- 3rd tarsal segment not concealed in this way J7

J7. Head extended in front of eyes in the form of a rostrum (beak) longer than the width of the head J8

- Head without a rostrum of this length J10

J8. Scape (lengthened basal segment of antenna) longer than the next 2 segments together .... **CURCULIONIDAE**

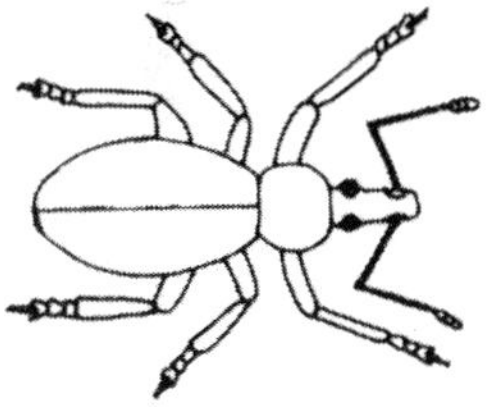

- Scape shorter than the next 2 segments together J9

J9. Scape nearly equal to the next 2 segments together .... **APIONIDAE**

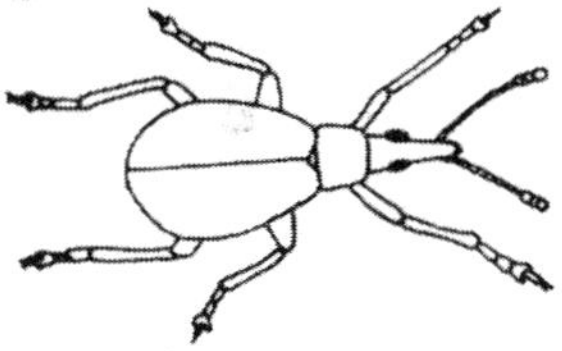

- Scape very short .... **ATTELABIDAE**

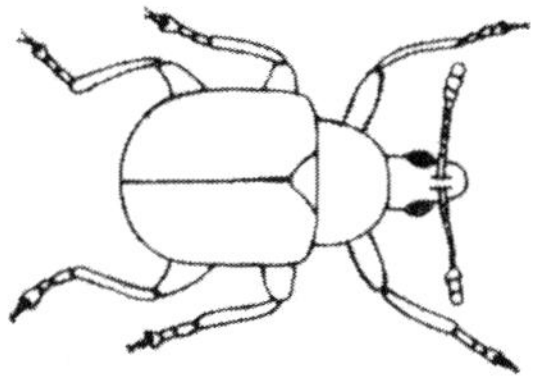

Note: Some Nemonychidae, G16, may key out here if antennae are thought to be clubbed. They have a long thin rostrum.

J10. Antennae consisting mainly of a very long scape (lengthened basal segment), and a long club; the whole antenna consisting of 3-7 segments total .... **SCOLYTIDAE**

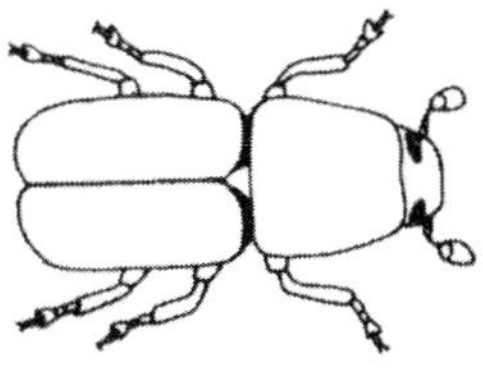

- Antenna with the middle section longer; the whole antenna composed of 8 segments or more J11

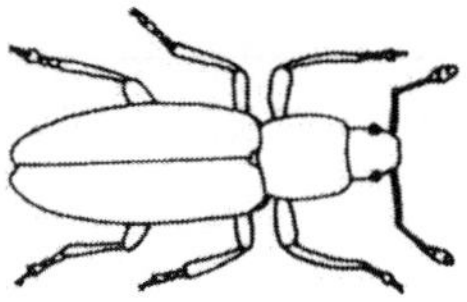

J11. Scape longer than the next 2 segments together, and quite thin basally .... **CURCULIONIDAE**

- Scape usually shorter than the next 2 segments, but if as long, then much fatter than segments 3-5 (or no obvious scape) J12

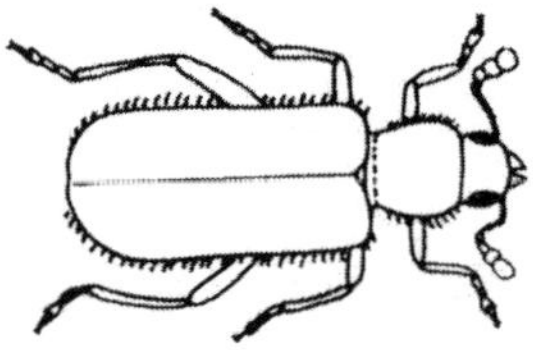

J12. Beetles over 5 mm long with long outstanding hairs. Often brightly coloured .... **CLERIDAE**

- Beetles bare, or with flattened pubescence J13

J13. Antennae with a very long club of 5-6 segments (could almost be described as an expanded antenna) .... **CHRYSOMELIDAE** (see G20)

- Antennae with a more obvious club, usually of 3 segments J14

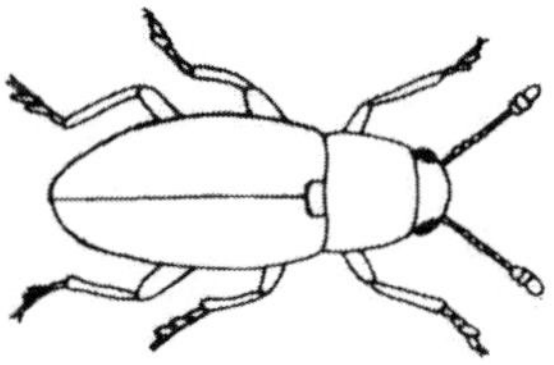

J14. Thorax as long as broad; sides not serrated, front corners not produced forwards .... **CRYPTOPHAGIDAE**

- Thorax in length at most 2/3 width; if longer, then the sides serrated or front corners produced forwards J15

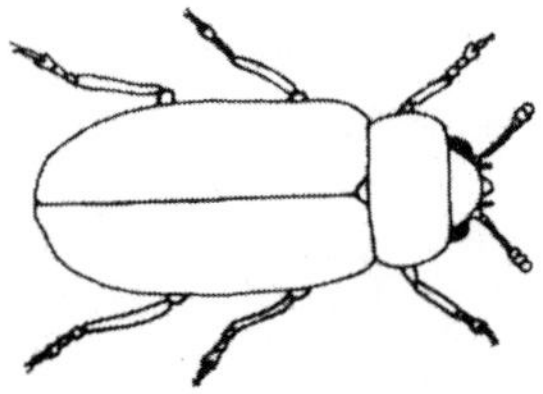

J15. Whole beetle deep yellow ground colour covered in fine yellow pubescence .... **BYTURIDAE**

- Not fitting this description J16

J16. Beetles with any of these characters: a) Elytra truncate, exposing the end of the abdomen; b) Tibiae flattened and expanded, toothed or serrated on outer edge; c) Antennae with large spherical club and swollen 1st (basal) segment .... **NITIDULIDAE**

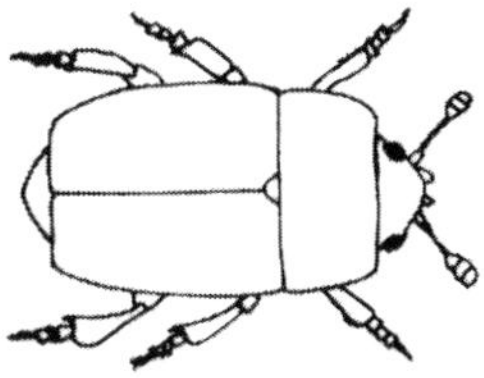

- Beetles without any of these characters J17

J17. Oval, hairless, domed beetles, flat underneath .... **PHALACRIDAE**

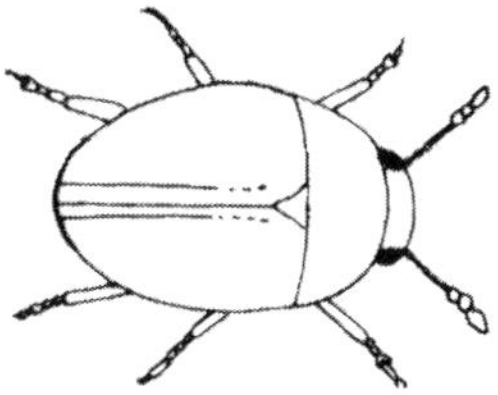

- More elongate beetles, convex underneath J18

J18. Thorax shorter than wide, without serrated sides ... **EROTYLIDAE**

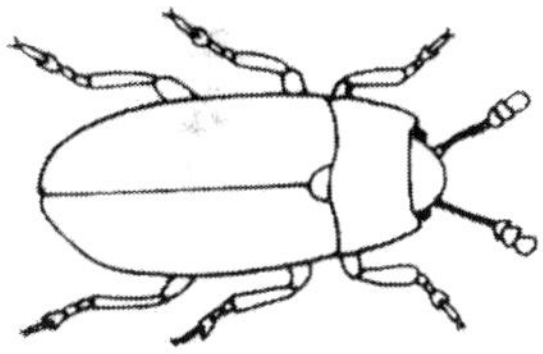

- Thorax as long as wide, with front corners produced or having serrated sides .... **SILVANIDAE**

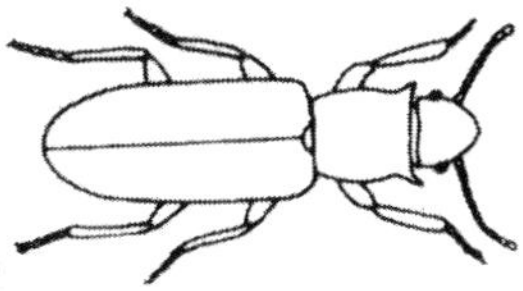

J19. Antennae much longer than the width of the thorax .... **ENDOMYCHIDAE**

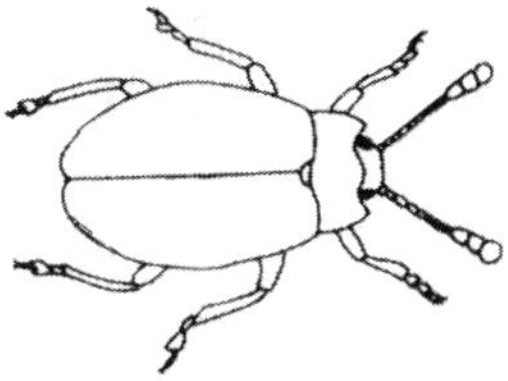

- Antennae at most as long as the width of the thorax, and usually much shorter .... **COCCINELLIDAE**

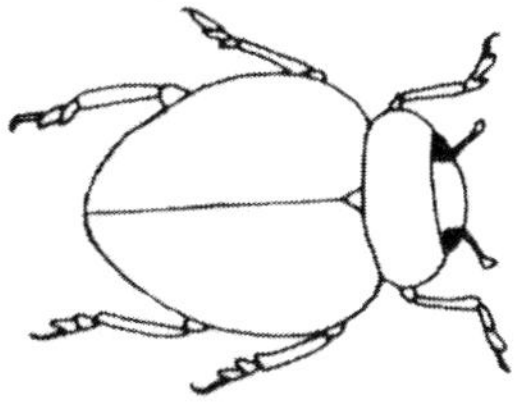

Key K

Elytra covering most of the abdomen. Antennae with a distinct club, and hind tarsi without lobed segments.

K1. Beetles under 1.2 mm long (excluding antennae) K2

- Beetles at least 1.2 mm long K6

K2. Beetles with characteristic antennae, having the first 2 segments stout, segment 3 onwards very thin, and a loose 3-segmented club .... **PTILIIDAE**

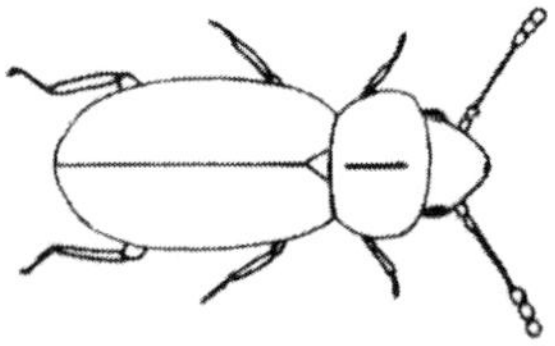

- Antennae not like this K3

K3. Oval, smooth, domed beetles, flat underneath, without long hairs or punctures on thorax or elytra (may be finely pubescent) K4

- Beetles without these characters K6

K4. Head very small, covered or nearly covered by thorax ..... **CORYLOPHIDAE**

- Head larger, not covered by thorax K5

K5. Head very wide, almost as wide as the thorax K6

- Head obviously narrower than the thorax .... **SPHAERIIDAE**

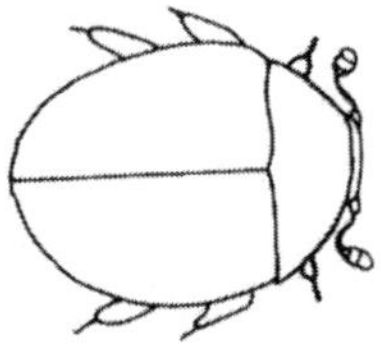

K6. Hind coxae very large, plate-like, covering about half of the underside of the abdomen (see F5). Head very wide .... **CLAMBIDAE**

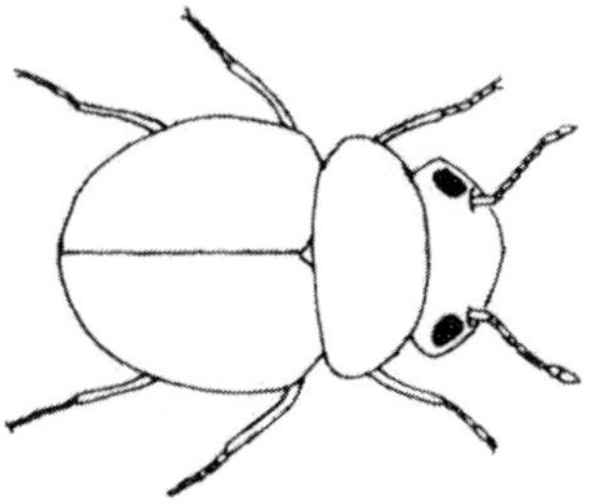

- Hind coxae much smaller than this K7

K7. Palps at least 2/3 length of antennae F3

Note: Antennae have more than 3 segments, palps never more than 3.

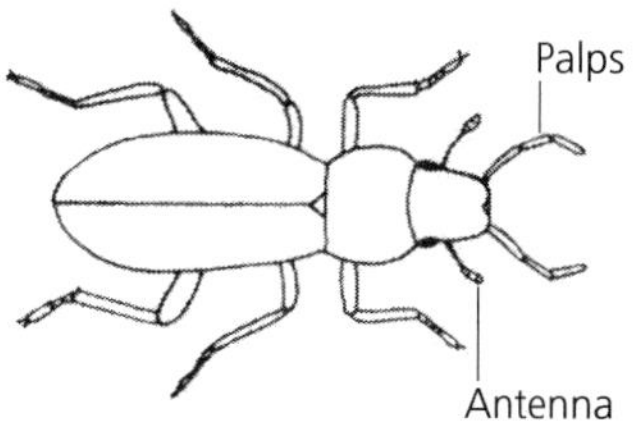

Palps

Antenna

\- Palps shorter than 2/3 length of antennae K8

K8. Thorax with a very distinct ridge down both sides, parallel to the edge, surmounted by a row of bent bristles .... **BIPHYLLIDAE**

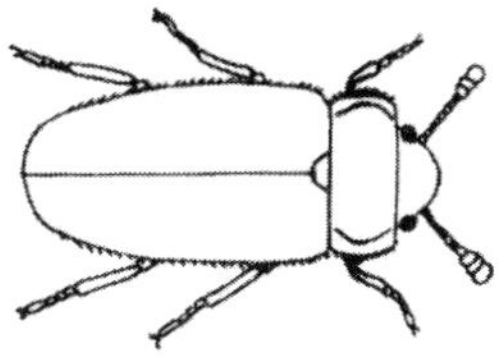

\- Thorax without these features K9

K9. Thorax shaped like a hood, concealing head from above, and covered with large bumps .... **BOSTRICHIDAE**

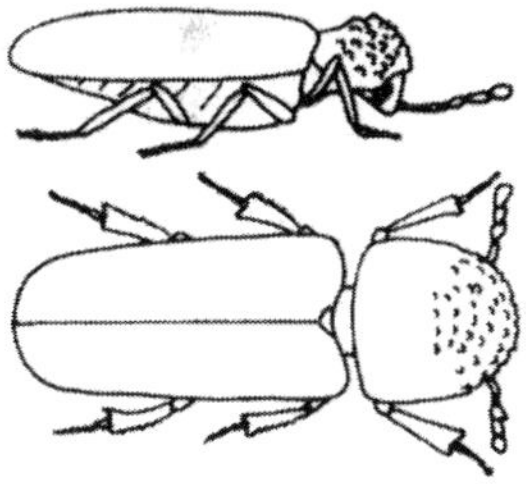

\- Thorax not hood-shaped, or without large bumps K10

K10. Oval beetles, length (excluding antennae) not more than twice the width K11

\- More elongate beetles K16

K11. Elytra truncate, exposing the end of the abdomen (look carefully at the back of the beetle, rather than from above) K12

\- Elytra not truncate, covering all of the abdomen K14

K12. Elytra each with exactly 9 longitudinal rows of fine punctures ... . **SPHAERITIDAE**

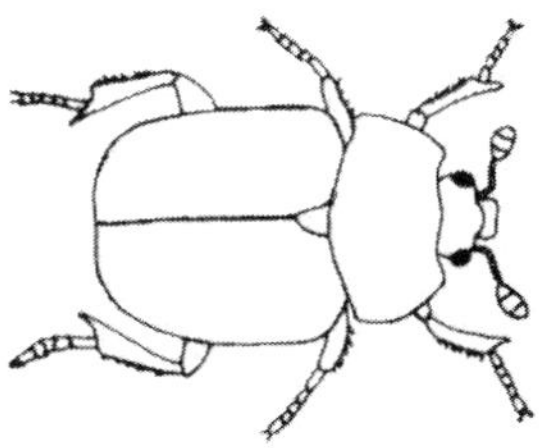

\- Beetles with other puncturation K13

K13. Antennae with scape (basal segment) about equal to next 2 segments together; club of antennae roughly circular .... **HISTERIDAE**

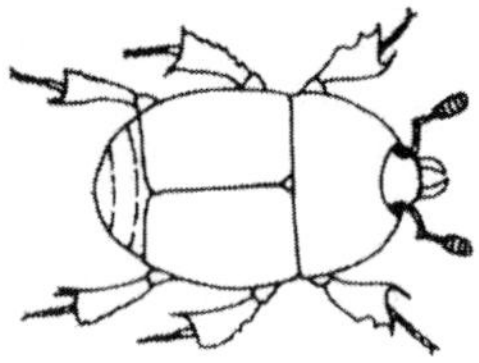

Note: Check the tarsi carefully once again; if they have bilobed segments try Nitulidae, see J16.

- Scape shorter than next 2 segments together; antennal club long .... **SCAPHIDIIDAE**

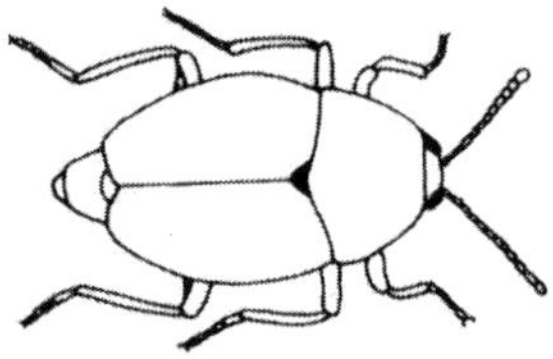

K14. Beetles with the underside very convex, with grooves to accommodate the legs when retracted K15

- Beetles flat underneath, or without grooves to accommodate legs K16

K15. Tibiae flattened, expanded and grooved to accommodate tarsi when folded .... **BYRRHIDAE**

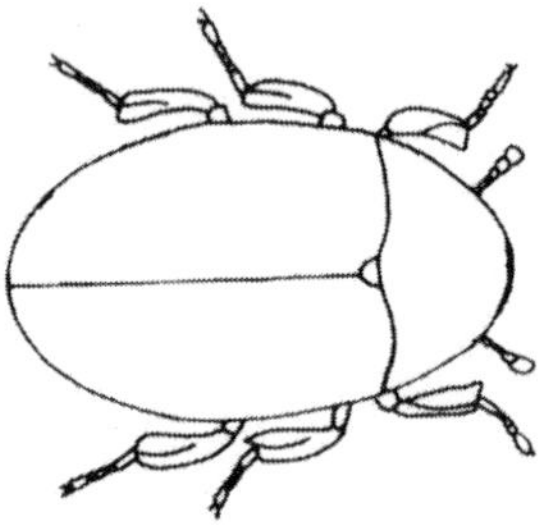

- Tibiae slim, without grooves .... **LIMNICHIDAE**

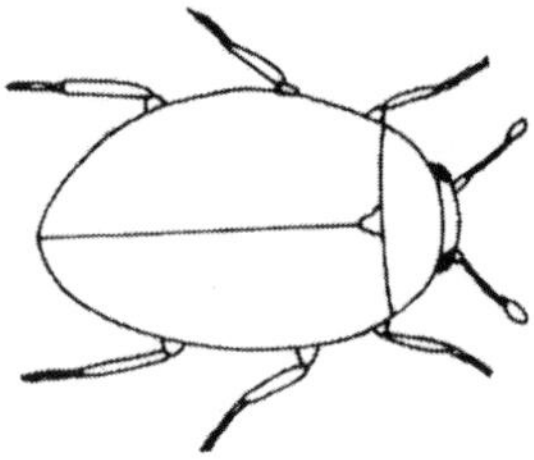

K16. Tarsi very long, over twice as long as tibiae; elongate, cylindrical beetles .... **PLATYPODIDAE**

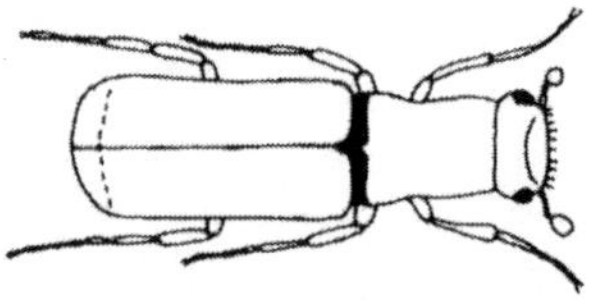

- Tarsi less than twice as long as tibiae K17

K17. Number of tarsal segments:

Note: Tarsal characters are often easier to see, particularly in small beetles, if a drop of water is placed on the tarsi.

Claws do not count as segments.

Legs: front	middle	hind	
5	5	5	.... K18
5	5	4	.... K27
5	4	4	.... K29
4	4	4	.... K32
3	4	4	.... K32
5	3	3	.... **LEIODIDAE**
4	3	3	.... **LEIODIDAE**
3	3	3	.... K46

K18. Beetles over 8 mm long (excluding antennae) .... **SILPHIDAE**

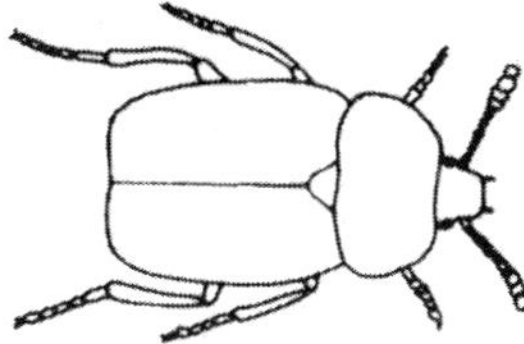

Note: Some Scarabaeidae may key out here if the lamellate nature of the antennae has been missed. They have the outer edges of the tibiae serrate.

- Beetles not over 8 mm long K19

K19. Thorax distinctly longer than broad, rectangular or wider in front. Elytra parallel-sided and completely covering the abdomen ....

.... **LYCTIDAE**

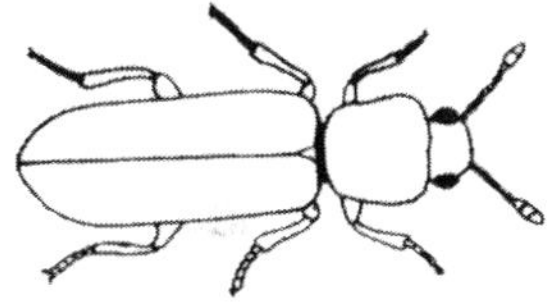

Note: If elytra are truncate, and antennae have a spherical club, try Rhizophagidae, K27. If elytra are very obviously rounded, go to K22.

- Thorax not longer than wide K20

K20. Thorax about half as long as wide, with broad side-margins. (Width of side-margin at base about equal to diameter of tibia at apex.) 3-segmented antennal club .... **PHLOIOPHILIDAE**

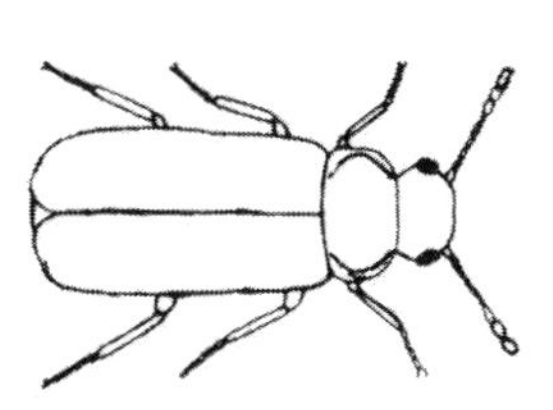

- Thorax longer than this, or without side margins K21

K21. Hind corners of the thorax extended rearwards into sharp points. Prosternal extension (see Fig. 3, p. 3) long; underside with short pubescence .... **THROSCIDAE**

- Hind corners of the thorax not pointed. Prosternal extension short K22

K22. Thorax clearly narrower than the elytra; small brownish beetles with scattered fine hairs .... **SCYDMAENIDAE**

- Thorax about as wide as the elytra K23

K23. Thorax with indentations near hind corners, or elytra with indentations near front corners .... **SCYDMAENIDAE**

- Thorax and elytra without these indentations K24

K24. Antennal club with up to 3 segments K25

- Antennal club long, with more than 3 segments. Segment 8 shorter or narrower than 7 or 9. Thorax always rounded in front .... **LEIODIDAE**

K25. Tibiae expanded and hooked, or with large spurs, or front tarsi lobed .... **LEIODIDAE**

Note: Some Nitidulidae may key out here, J16; they have an almost spherical antennal club, composed of three segments.

- Tibiae and tarsi simple K26

K26. Femora short enough to be folded away under edge of elytra: hind coxae (see Fig. 3, p. 3) excavated on posterior face to accommodate femora .... **DERMESTIDAE**

Note: If antennal club consists of 3 almost triangular segments, it may be Anobiidae, H14; they have much longer antennae, about as long as front legs.

- Femora longer, too long to be folded under elytra: hind coxae not excavated as above .... **CRYPTOPHAGIDAE**

K27. Elytra truncate, exposing end of abdomen. Beetles at least 1.4 mm long (excluding antennae). (If under 1.4 mm, try Hypocopridae, K33.) .... **RHIZOPHAGIDAE**

- Elytra not truncate, usually covering all of the abdomen K28

K28. Antennae with a very obvious 4-segmented club, almost half as long as the complete antenna. Head and elytra black, thorax yellow. Tibiae slender .... **TETRATOMIDAE**

Note: If tibiae are expanded, with spurs or serrations on the outside, try Leiodidae, K25.

- Antennae with club not like this K29

K29. Antennae with a spherical club .... **COLYDIIDAE**

- Antennal club elongate K30

K30. Eyes higher than wide or notched in front .... **TENEBRIONIDAE**

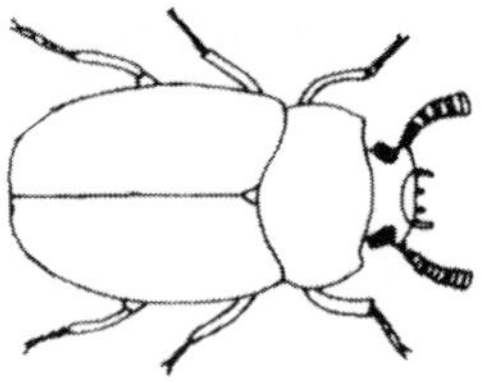

- Eyes approximately circular K31

K31. Antennae with a 3-segmented club, the last segment long and conical .... **DERMESTIDAE**

- Antennal club not like this K24

K32. Beetles less than 1.4 mm long (excluding antennae). Elytra truncate, exposing back of abdomen .... **HYPOCOPRIDAE**

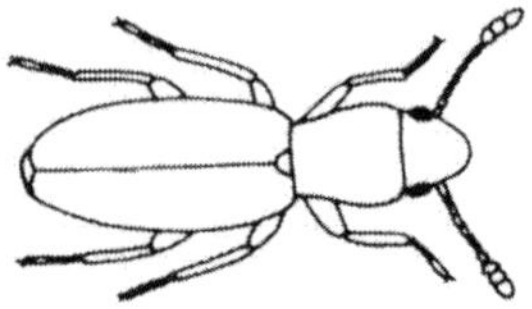

Note: If over 1.4 mm, try Rhizophagidae, K27.

- Beetles at least 1.4 mm long, or elytra completely covering abdomen K33

K33. Thorax with a wide turned-out rim, giving a terrapin-like appearance .... **PELTIDAE**

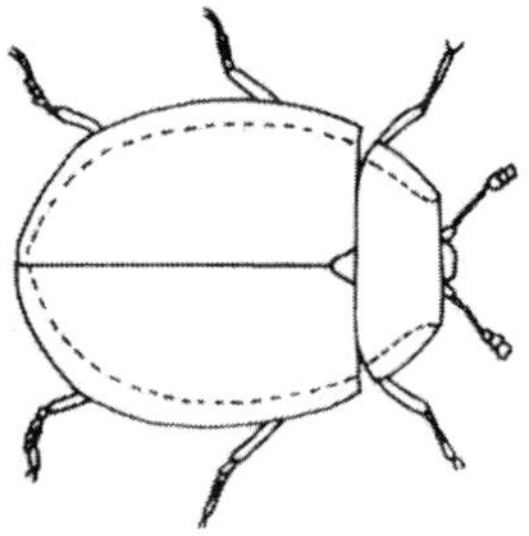

- Thorax and elytra without this kind of margin K34

K34. Antennae with fewer than 11 segments, including a club of 2 segments or more K35

- Antennae with 11 segments, or with a spherical club of 1 segment K36

K35. Punctures on elytra in straight, parallel rows .... **SPHINDIDAE**

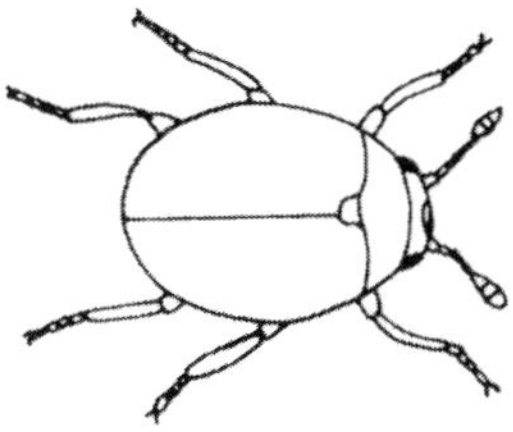

- Punctures on elytra always randomly distributed on the forward part; sometimes in rows at the apex .... **CISIDAE**

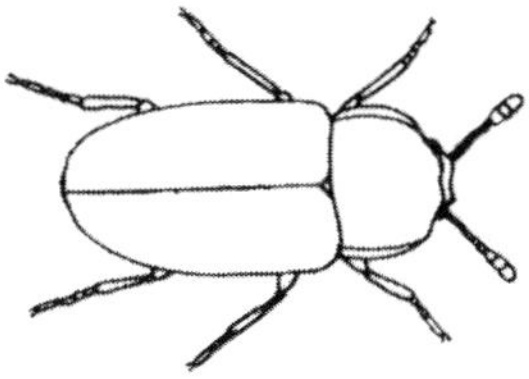

K36. Thorax with outstanding hairs. Club of antennae with 3 segments .... **ENDOMYCHIDAE**

Note: If a weak club of about 5 segments, try Mycetophagidae, K43.

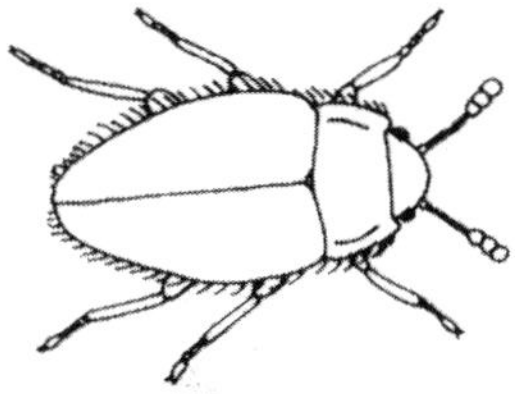

- Thorax with flattened pubescence at most K37

K37. Antennae with a spherical club composed of 1 segment, sometimes with 1 or 2 vestigial segments apically K38

- Antennae with a club of 3 roughly equal segments or more K39

K38. Shining, lightly punctured, smooth hairless beetles .... **CERYLONIDAE**

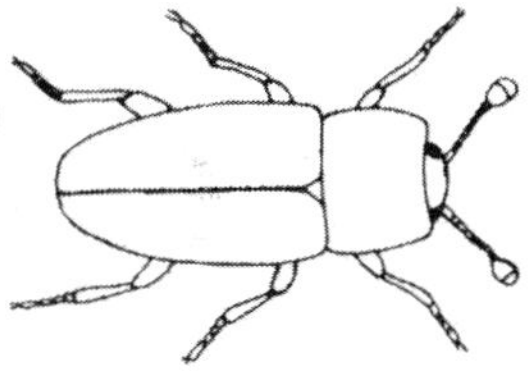

- Quite dull, densely punctured beetles, covered in microscopic flattened hairs .... **COLYDIIDAE**

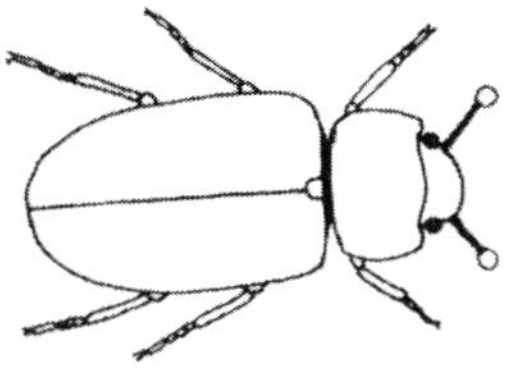

K39. Antennae with an almost spherical club composed of 3 roughly equal segments .... **NITIDULIDAE** (see J16)

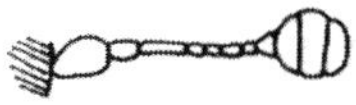

- Club of antennae more extended K40

K40. Head very long, cleft in front; eyes well in front of the thorax .... **TROGOSSITIDAE**

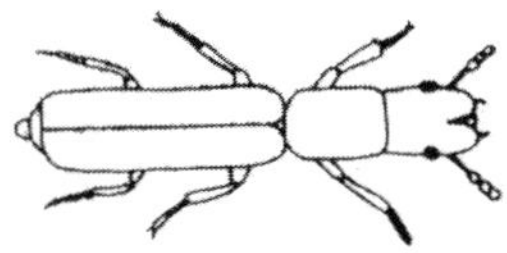

- Head shorter, without these characters K41

K41. Antennae very hairy .... **COLYDIIDAE**

- Antennae without such hairs K42

K42. Antennal club with more than 3 segments K43

- Antennal club with 3 segments only K44

K43. Thorax as wide or wider in front as at the rear. Uniformly brown beetles .... **COLYDIIDAE**

- Thorax narrower in front than at the rear. Brown beetles with orange markings .... **MYCETOPHAGIDAE**

  Note: Some Leiodidae, K24, may key out here; they have antennal segment 8 reduced compared with 7 or 9.

K44. Rather flattened beetles: elytra parallel-sided, brownish orange, hairless, lightly stippled with microscopic punctures. Not striated .... **CUCUJIDAE** (see H25)

- Elytra usually with rounded sides, but if parallel-sided, then clearly striated K45

K45. Thorax broader than long .... **MYCETOPHAGIDAE** (see K43)

- Thorax longer than broad with sharp front corners, striated or roughly sculptured .... **COLYDIIDAE** (see K47)

K46. Antennae with 8-10 segments .... **CISIDAE**

- Antennae with 11 segments K47

K47. Thorax much longer than broad .... **COLYDIIDAE**

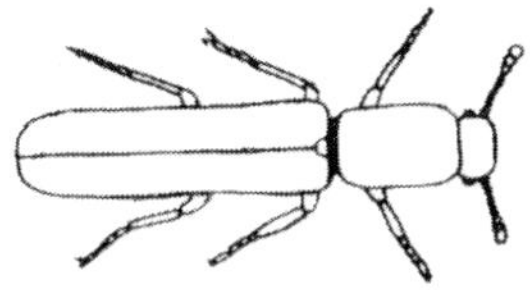

- Thorax at most as long as broad K48

K48. Club of antennae almost spherical .... **CERYLONIDAE**

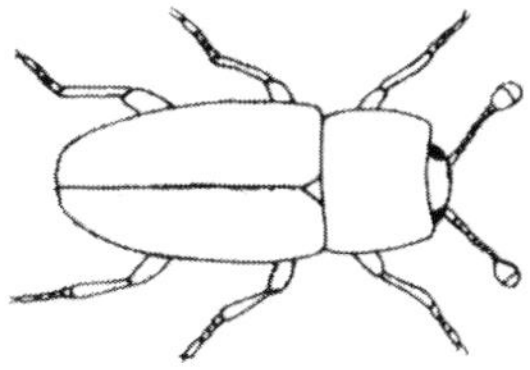

- Club of antennae more elongate K49

K49. Upper surface smooth, without hairs .... **MEROPHYSIIDAE**

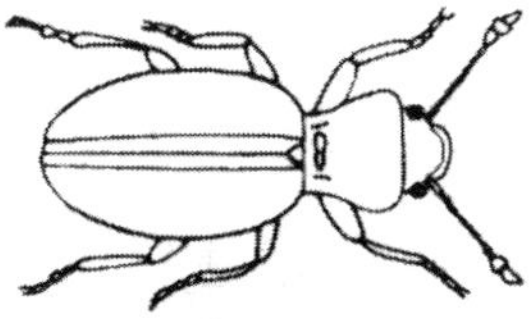

- Upper surface wrinkled, clearly punctured or pubescent .... **LATHRIDIIDAE**

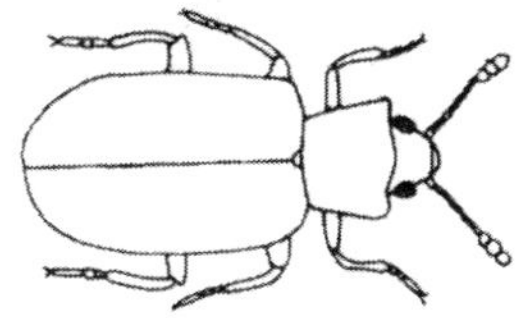

Key L

ORDER STREPSIPTERA

Elytra reduced to knob-like structures; hindwings large and fully developed. Females not free-living, but in a puparium inside the host (Hymenoptera or Hemiptera).

L1. Tarsi with 2 segments. Antennae with 4 segments, the 3rd with side extension .... **ELENCHIDAE**

- Tarsi with 3 segments. Antennae with 7 segments, 3rd-6th with side extensions .... **HALICTOPHAGIDAE**

- Tarsi with 4 segments. Antennae with 6 segments, 3rd with side extension .... **STYLOPIDAE**

Alphabetical list of families of British Coleoptera

Names follow Pope (1977). Size of beetles does not include antennae. Tarsal segments are given for each leg, starting from the front legs, and ignoring small cylindrical segments immediately following bilobed segments. References to Joy mean Joy (1932) and those to Lyneborg mean Lyneborg (1977).

ADERIDAE. 1 g. 3 spp. 1.5-2.4 mm. Tarsi 4,4,4 (may look 4,4,3) with some weakly bilobed segments. Antennae filiform. Very finely pubescent, with large coarsely granulated eyes which are notched in front to accommodate the antennae. Entirely reddish, or with head and thorax black. Found in or near old wood. Keys out at G1 and H2. RES Handbook by Buck (1954). Illustrated by Joy, Plate 88 (4 & 5).

ANOBIIDAE. 17 g. 27 spp. 1.3-7 mm. Tarsi 5,5,5 simple. Antennae filiform, serrate, pectinate, or clubbed. This family includes woodworm and death watch beetles. Found in wood and stored products. Keys out at E6, H14 & K26. Illustrated by Joy, Plate 131 (I-II, 14-15) and by Lyneborg, Plate 224-227.

ANTHICIDAE. 2 g. 16 spp. 2.0-4.5 mm. Tarsi 5,5,4 with some bilobed segments. Antennae filiform. *Notoxus* has the thorax extended over the head in the form of a horn. Reddish yellow with a black pattern on the elytra. Found in sandy ground including riverbanks. *Anthicus* has no pronotal horn, and is brownish-black. It is found around compost and manure heaps and in sandy places. Keys out at G13 & G14. RES Handbook by Buck (1954). Illustrated by Joy, Plate 88 (6-13).

ANTHRIBIDAE. 6 g. 8 spp. 1.4-13.0 mm. Tarsi 4,4,4 with some bilobed segments. Antennae filiform or clubbed. Related to the weevils, but rostrum wide and short and without a lengthened 1st antennal segment. Associated with rotting wood and fungi. Keys out at G17 & J6. Illustrated by Joy, Plate 68 (7-10) and Lyneborg, Plate 395.

APIONIDAE. 2 g. 73 spp. 1.1-4.4 mm. Tarsi 4,4,4 with some bilobed segments. Related to the weevils, with a long rostrum, but antennae clubbed with a short 1st segment. Associated with shrubs and herbs. Keys out at J9. Illustrated by Joy, Plate 46-47 and by Lyneborg, Plate 400-403.

ATTELABIDAE. 5 g. 21 spp. 4.0-8.0 mm. Tarsi 4,4,4 with some segments bilobed. Antennae clubbed with very short 1st segment. Rostrum robust, short or long. Larvae in leaf rolls in fruit trees, hawthorn and hazel. Keys out at J5 & J9. Illustrated by Joy, Plate 45 and Lyneborg, Plate 396-399.

BIPHYLLIDAE. 2 g. 2 spp. 3-3.5 mm. Tarsi 5,5,5 simple. Antennae clubbed. Dark pubescent beetles with ridges on thorax. Found in fungi on bark. Keys out at K8. Illustrated by Joy, Plate 155 (1).

BOSTRICHIDAE. 3 g. 3 spp. 2.5-9 mm. Tarsi 5,5,5 simple. Antennae clubbed. Cylindrical beetles with the front part of the thorax covered in bumps. Larvae make tunnels in fallen timber. Keys out at K9. Illustrated by Joy, Plate 131 (12-13) and Lyneborg, Plate 230.

BRUCHIDAE. 5 g. 13 spp. 2-5 mm. Tarsi 4,4,4 with some bilobed segments. Antennae serrate or filiform. Rostrum short, elytra with mottled pattern. Associated with plants in the pea family; *Acanthophilus* is known as the bean weevil. Keys out at G19. Illustrated by Joy, Plate 114 (6-8) and Lyneborg, Plate 393-394.

BUPRESTIDAE. 5 g. 12 spp. 1.7-13 mm. Tarsi 5,5,5 with some bilobed segments. Antennae filiform, serrate or weakly clubbed. Often elongate, pointed at rear, with bright metallic coloration. Adults found in flowers and on tree trunks; larvae live in wood or in roots. Keys out at E5, G2 & J3. RES Handbook by Levy (1977). Illustrated by Joy, Plate 129 and Lyneborg, Plate 166, 172 & 173.

BYRRHIDAE. 6 g. 12 spp. 1.2-10 mm. Tarsi 5,5,5 simple or 4,4,4 simple. Antennae clubbed. Short oval beetles which can fold up their legs under the body. Larvae live in the ground; adults found under stones, in moss, etc. Keys out at K15. Illustrated by Joy, Plate 137 (1,3 & 4) and Lyneborg, Plate 149-150.

BYTURIDAE. 1 g. 2 spp. 3-4.5 mm. Tarsi 4,4,4 with some bilobed segments. Antennae clubbed. Small, cylindrical brown or grey beetles covered with silky hairs. Associated with flowers, whitethorn, strawberry, etc. Keys out at J15. Illustrated by Joy, Plate 163 (5) and Lyneborg, Plate 263-264.

CANTHARIDAE. 6 g. 41 spp. 1-15 mm. Tarsi 5,5,5 with some bilobed segments. Antennae filiform. Elongate beetles with parallel-sided soft elytra. Long and slender legs and antennae. Adults seen mostly on flowers; larvae in the ground, often in moss. This family includes the soldier beetles. Keys out at B9, G8 & G9. Illustrated by Joy, Plate 124 (6), 125 & 126 (4-6), and Lyneborg, Plate 195-200.

CARABIDAE. 62 g. 342 spp. 2-30 mm. Tarsi 5,5,5 mostly simple. Antennae filiform. In the Carabinae (ground beetles), head is small, thorax broader than head and elytra long-oval. Antennae inserted at side of head between eyes and mandibles. In the Cicindelidae (tiger beetles), head is larger, with protruding eyes. Antennae inserted above base of mandibles. Carabidae are mostly predatory. Keys out at H3. RES Handbook by Lindroth (1974). Illustrated by Joy, Plate 93-109 and Lyneborg, Plate 1-44.

CERAMBYCIDAE. 37 g. 65 spp. 3.5-40.0 mm. Tarsi 4,4,4 with some bilobed segments. Antennae long, filiform. Mostly oblong beetles with long antennae. Some have cryptic coloration, some brightly coloured. Most species are associated with trees, some being serious timber pests. Longhorn beetles. Larvae under bark or in plant stems. Most have eyes notched in front to accommodate antennae. Keys out at B9, G19 & G20. RES Handbook by Duffy (1952). Illustrated by Joy, Plate 110-113 and Lyneborg, Plate 296-352.

CERYLONIDAE. 3 g. 6 spp. 0.8-2.5 mm. Tarsi 3,3,3 or 4,4,4 simple. Antennae clubbed. Small beetles with 1- or 2-segmented spherical antennae club and last tarsal segment often longer than the rest together, or side of thorax deeply notched in front. Keys out at K38 & K48. Illustrated by Joy, Plate 148 (3).

CHRYSOMELIDAE. 52 g. 254 spp. 1.0-18.0 mm. Tarsi 4,4,4 mostly with some bilobed segments (if simple, the last tarsal segment very long). Antennae filiform serrate or expanded. Often broad-oval, brightly coloured beetles. Some resemble ladybirds, but never have clubbed antennae. Often found on leaves, this family includes the flea beetle, the Colorado beetle and other pest species. Keys out at G20, H22, J11 & J13. Illustrated by Joy, Plate 114 (1-5, 9-15) & Plate 115-121 and Lyneborg, Plate 353-392.

CISIDAE. 4 g. 22 spp. 1.2-3.0 mm. Tarsi 4,4,4 or 3,3,3 simple. Antennae clubbed. Small beetles with 8-10 antennal segments. Found under bark of trees, and in tree-related fungi such as *Polyporus*. Keys out at K35 & K46. Illustrated by Joy, Plate 162 (3-15).

CLAMBIDAE. 2 g. 9 spp. 0.9-1.8 mm. Tarsi 5,5,5 simple. Antennae clubbed. Very small, oval beetles with very wide heads and large plate-like hind coxae. Associated with decaying vegetable matter. Keys out at K6. RES Handbook by Johnson (1966). Illustrated by Joy, Plate 136 (13-15).

CLERIDAE. 10 g. 14 spp. 3-10 mm. Tarsi 5,5,5 or 4,4,4 always with some bilobed segments. Antennae expanded or clubbed. Medium-sized beetles, often brightly coloured and patterned with outstanding hairs. Larvae predatory, under bark or in carrion. Keys out at G6, J2 & J12. Illustrated by Joy, Plate 123 (2-7).

COCCINELLIDAE. 27 g. 42 spp. 1.0-9.0 mm. Tarsi 3,3,3 with some segments bilobed. Antennae clubbed. Mostly broad, oval beetles with bright coloration, often patterned with spots. Ladybird beetles, which prey on aphids (greenfly). Keys out at J19. RES Handbook by Pope (1953). Illustrated by Joy, Plate 149 (3)-152 and Lyneborg Plate 251-262.

COLYDIIDAE. 13 g. 17 spp. 1.3-6.0 mm. Tarsi 4,4,4 or 3,3,3 simple (rarely 5,4,4). Antennae clubbed. Small beetles living under bark or in fungi. Keys out at K29, K38, K41, K43, K45 & K47. Illustrated by Joy, Plate 148 (1,2,5-8) & Plate 145 (4-5). and Lyneborg, Plate 248.

CORYLOPHIDAE. 4 g. 10 spp. 0.5-1.0 mm. Tarsi simple. Antennae clubbed. Very small, oval, domed beetles; hairless and flat underneath. Found in decaying vegetation and fungi. Keys out at K4. Illustrated by Joy, Plate 169.

CRYPTOPHAGIDAE. 10 g. 110 spp. 1.3-11 mm. Tarsi 5,5,5 or 5,5,4 simple, or 4,4,4 with some bilobed segments. Antennae clubbed. Oval or long-oval, yellow brown or black without any pattern; often pubescent. Found in decaying vegetation and fungi. Keys out at J14 & K26. Illustrated by Joy, Plate 143 (4-12) & Plate 144, and Lyneborg, Plate 242-243.

CUCUJIDAE. 7 g. 14 spp. 3.5-4.5 mm. Tarsi 5,5,5, 5,5,4 or 4,4,4 simple. Antennae filiform or clubbed. Small to medium-sized, rather flattened beetles with squarish thorax, often wider in front than at base. Beetles predatory under bark. Keys out at H1, H25 & K44. Illustrated by Joy, Plate 137 (7,9-11) and Lyneborg, Plate 239-241.

CURCULIONIDAE. 102 g. 416 spp. 1.5-14.0 mm. Tarsi 4,4,4 with some bilobed segments. Antennae clubbed, with scape. The weevils have a long rostrum (snout) and/or elbowed antennae. Most are covered with small scales. All are herbivorous and many are agricultural pests. Keys out at J8 & J11. Illustrated by Joy, Plate 46-65 and Lyneborg, Plate 404-432.

DASCILLIDAE. 1 g. 1 sp. 9-11 mm. Tarsi 5,5,5 with some bilobed segments. Antennae filiform. Brownish-black with light brown pubescence. Thorax with obvious side-margins. Adults seen mostly in flowers. Keys out at G8. Illustrated by Joy, Plate 123 (8) and by Lyneborg, Plate 146.

DERMESTIDAE. 9 g. 30 spp. 1.5-12 mm. Tarsi 5,5,5 simple (may look 5,5,4). Antennae with club. Very robustly built beetles, often covered with fine hair or scales. Larvae feed on animal remains (skins, etc.). This family includes the museum beetles. Keys out at K26 & K31. Illustrated by Joy, Plate 141 (3-9) and Lyneborg, Plate 219-223.

DRILIDAE. 1 g. 1 sp. 5-7 mm. Tarsi 5,5,5 with some bilobed segments. Male antennae pectinate. Elongate beetles; males with orange-brown elytra covered with pale hairs. Female antennae filiform; elytra absent. Snail predators. Keys out at B4 & E5. Illustrated by Joy, Plate 123 (1).

DRYOPIDAE. 2 g. 8 spp. 3.5-5.5 mm. Tarsi 5,5,5 simple. Antennae unusual. Small beetles with short antennae and covered in fine hairs. Found close to fresh water; larvae aquatic. Keys out at E2. Illustrated by Joy, Plate 142 (1-2) and Lyneborg, Plate 152.

DYTISCIDAE. 27 g. 113 spp. 1.7-38 mm. Tarsi 5,5,5. Hind tarsi with swimming hairs. Antennae filiform. Boat-shaped predatory water beetles. Keys out at F7, F8 & F9. RES Handbook by Balfour-Brown (1953). Illustrated by Joy, Plate 77 & 78 (1-14, 16-19), Plate 79-81 and Lyneborg, Plate 47-48, & 50-57.

ELATERIDAE & **EUCNEMIDAE**. 29 g. 68 spp., and 5 g. 6 spp respectively. 2.8-30 mm. Tarsi 5,5,5 simple. Antennae filiform, serrate or pectinate. These are the 'click' beetles. Elateridae have the labrum (upper plate just over the mandibles) protruding forwards of the front of the head, while in Eucnemidae the labrum is not visible from above. Adults found on vegetation, larvae in wood or soil (wireworms). Keys out at E6 & H11. Illustrated by Joy, Plate 127 (2-16), Plate 128 and Lyneborg, Plate 176-191.

ELMIDAE. 8 g. 11 spp. 1.2-4 mm. Tarsi 5,5,5 simple. Antennae filiform. Small aquatic beetles with long legs, found under stones, or in moss. Keys out at H13. Illustrated by Joy, Plate 135 (4-11) and Lyneborg, Plate 154.

ENDOMYCHIDAE. 5 g. 6 spp. 1.0-5.0 mm. Tarsi 4,4,4 simple or 3,3,3 with some segments bilobed. Antennae clubbed or filiform. Small, oval, or long-oval beetles; hairless, or with scattered hairs. Associated with fungi. Keys out at G1, J19 & K36. Illustrated by Joy, Plate 141 (1-2) and Lyneborg, Plate 249-250.

EROTYLIDAE. 3 g. 7 spp. 2.5-7 mm. Tarsi 5,5,5 or 4,4,4 with some bilobed segments. Antennae clubbed. Oval, hairless beetles associated with fungi on trees. Keys out at J4 & J18. Illustrated by Joy, Plate 155 (7-9) and Lyneborg, Plate 244.

EUCINETIDAE. 1 g. 1 sp. 3.2-4 mm. Tarsi 5,5,5 with front tarsi lobed. Antennae filiform. Oval pubescent beetles, pointed in the rear. Keys out at H7.

EUCNEMIDAE. see Elateridae above.

GEOTRUPIDAE. 3 g. 8 spp. 12-25 mm. Tarsi 5,5,5 simple. Antennae with lamellate club. Large, generally black and iridescent dung beetles. Eggs laid in dung, in which larvae feed. Keys out at D4. RES Handbook by Britton (1956). Illustrated by Joy, Plate 70 (5-7) and Lyneborg, Plate 107-111.

GYRINIDAE. 3 g. 13 spp. 4.5-7.5 mm. Tarsi unusually modified for aquatic use. Antennae short and unusual. Whirligig beetles; oval beetles that form flotillas on the water surface. Keys out at F1. Illustrated by Joy, Plate 74 and Lyneborg, Plate 58-59.

HALIPLIDAE. 3 g. 18 spp. 2-4.0 mm. Tarsi 5,5,5. Hind tarsi with swimming hairs. Small boat-shaped aquatic beetles with unusually large plate-like hind coxae. Adults feed on green algae and are found in water with dense vegetation. Keys out at F5. RES Handbook by Balfour-Browne (1953). Illustrated by Joy, Plate 75 and Lyneborg, Plate 46.

HETEROCERIDAE. 1 g. 8 spp. 2.5-5.9 mm. Tarsi 4,4,4 simple. Unusual antennae. Oblong hairy beetles with sides of elytra parallel, large mouthparts and very unusual antennae. They live in galleries in banks adjacent to fresh water or salt water, and in soil. Keys out at E2. RES Handbook by Clarke (1973). Illustrated by Joy, Plate 135 (1-3) and Lyneborg, Plate 155.

HISTERIDAE. 22 g. 49 spp. 0.8-10 mm. Tarsi 5,5,5 simple. Antennae clubbed. Short oval beetles with truncate elytra and elbowed antennae. Adults and larvae predatory, found in rotting organic matter, carrion, dung, etc. Keys out at K13. RES Handbook by Halstead (1963). Illustrated by Joy, Plate 133 (1-10) and Lyneborg, Plate 71.

HYDRAENIDAE. 3 g. 29 spp. 1.1-4.5 mm. Tarsi 4,4,4 simple. Antennae clubbed. Aquatic beetles with long palps. Keys out at F4. Illustrated by Joy, Plate 86 (2-17) and Lyneborg, Plate 61.

HYDROPHILIDAE. 20 g. 89 spp. 1.2-60 mm. Tarsi 5,5,5 not bilobed. Antennae clubbed. Beetles with long palps, some being aquatic and found on the bottom among water plants; terrestrial forms in rotting vegetable matter. Keys out at F3 & F4. Illustrated by Joy, Plate 82-86 (1) and Lyneborg, Plate 63-70.

HYGROBIIDAE. 1 g. 1 sp. 8.5-10 mm. Tarsi 5,5,5 Hind tarsi with swimming hairs. Antennae filiform. Water beetles, strongly convex above and below. Known as screech beetles. Keys out at F7. Illustrated by Joy, Plate 76 (1).

HYPOCOPRIDAE. 1 g. 1 sp. 1-1.3 mm. Tarsi 4,4,4 simple. Antennae clubbed. Tiny dull-black beetles with truncate elytra. Found in cow dung. Keys out at K32. Illustrated by Joy, Plate 145 (6).

LAMPYRIDAE. 2 g. 2 spp. 5-16 mm. Tarsi 5,5,5 with some bilobed segments. Antennae filiform. Glow-worms. Females without elytra, males with elytra covering abdomen, or reduced. Predatory on snails. Keys out at B3, B8 & G4. Illustrated by Joy, Plate 122 (2) and Lyneborg, Plate 192.

LATHRIDIIDAE. 13 g. 61 spp. 1.0-3.0 mm. Tarsi 3,3,3 simple. Antennae clubbed (sometimes weakly). Thorax narrower than elytra, often sculptured. Found in decaying matter, or on vegetation. Keys out at H2 & K49. Illustrated by Joy, Plate 146 (1-6, 8-13) & 147 and Lyneborg, Plate 245-246.

LEIODIDAE. 20 g. 92 spp. 1.5-7 mm. Tarsi 5,5,5 or 5,5,4 or 5,3,3 or 4,3,3 simple. Antennae clubbed or filiform. Usually oval beetles, with variable tarsi but never with bilobed segments. Antennae in some species with a long club, including a reduced segment 8. Many species associated with fungi, and found under bark of trees. Others in vegetable matter. Keys out at H15, K17, K24, K25 & K28. Illustrated by Joy, Plate 138-140, 160-161 and Lyneborg, Plate 75.

LEPTINIDAE. 1 g. 1 sp. 2-2.5 mm. Tarsi 5,5,5 with some bilobed segments. Antennae filiform. Small beetles with long thin antennae and completely without eyes. Found in the burrows of rodents and birds nests. Keys out at G3. Illustrated by Joy, Plate 137 (5) and Lyneborg, Plate 74.

LIMNICHIDAE. 1 g. 1 sp. 1.5-1.8 mm. Tarsi 5,5,5 simple. Antennae clubbed. Very small beetles found in the margins of water courses. Antennae with 10 segments, weakly clubbed. Keys out at K15. Illustrated by Joy, Plate 137(2).

LUCANIDAE. 4 g. 4 spp. 10-70 mm. Tarsi 5,5,5 simple. Antennae with lamellate club. Large brownish or black beetles; males of some species with enormous mandibles. Stag beetles. Antennae elbowed. Larvae in rotting wood. Keys out at D1. RES Handbook by Britton (1956). Illustrated by Joy, Plate 69 and by Lyneborg, Plate 100 & 104.

LYCIDAE. 3 g. 4 spp. 5-9 mm. Tarsi 5,5,5 with some bilobed segments. Antennae filiform. Elongate beetles with soft elytra, sculptured thorax and long antennae. Found on umbelliferous plants. Larvae under bark and in rotting wood. Keys out at G5. Illustrated by Joy, Plate 122 (1) and Lyneborg, Plate 210.

LYCTIDAE. 2 g. 6 spp. 2.7-6 mm. Tarsi 5,5,5 simple. Antennae clubbed. Flat, elongate beetles with punctured striae on elytra, and antennae with 2-segmented club. Wood-boring beetles associated with felled oaks. Keys out at K19. Illustrated by Joy, Plate 145 (1-2).

LYMEXYLIDAE. 2 g. 2 spp. 6-16 mm. Tarsi 5,5,5 simple. Antennae filiform or serrate. Very long, slender beetles; elytra narrow towards rear, and end of abdomen exposed. Keys out at B7 and H5. Illustrated by Joy, Plate 124 (1-2) and Lyneborg, Plate 217-218.

MELANDRYIDAE. 11 g. 18 spp. 2.2-16.0 mm. Tarsi 5,5,4 with some bilobed segments or 5,5,4 or 4,4,4 simple. Antennae filiform. Somewhat elongate, narrow, hard beetles without much pubescence. Found in various habitats including whitethorn, fungal growth in old wood, and old willow stumps. Keys out at G12, G15, H19 & H26. RES Handbook by Buck (1954). Illustrated by Joy, Plate 92 (1, 3-8) and Lyneborg, Plate 276-277.

MELOIDAE. 3 g. 9 spp. 8.0-40.0 mm. Tarsi 5,5,4 simple. Antennae filiform. Medium to large beetles with a very abrupt 'neck'. Elytra complete or short. Found on open, grassy banks or on trees; larvae in bees' nests. Keys out at B7 & H21. RES Handbook by Buck (1954). Illustrated by Joy, Plate 88 (1-3) and Lyneborg, Plate 285-289.

MELYRIDAE. 10 g. 22 spp. 1.4-8 mm. Tarsi 5,5,5 simple (2nd segment of front tarsus sometimes lobed). Antennae filiform or serrate. Rather variable, often elongate beetles, with soft elytra, usually found on grasses, bushes or trees. Larvae live in timber under bark. Keys out at H8, H10 & H15. Illustrated by Joy, Plate 124 (3-5) and Lyneborg, Plate 206.

MEROPHYSIIDAE. 1 g. 3 spp. 1-1.7 mm. Tarsi 3,3,3 simple. Antennae clubbed. Small yellowish beetles with well marked sutural striae on elytra. Found in decaying vegetable matter. Keys out at K49. Illustrated by Joy, Plate 146 (7).

MORDELLIDAE. 3 g. 10 spp. 2.0-9.0 mm. Tarsi 5,5,4 simple. Antennae filiform. Tip of abdomen not covered by elytra. Adults seen on flowers and tree trunks. Larvae in rotting wood or in plant stems. Keys out at H17. RES Handbook by Buck (1954). Illustrated by Joy, Plate 89 (14-17) and Lyneborg, Plate 281-282.

MYCETOPHAGIDAE. 5 g. 12 spp. 2.0-6.0 mm. Tarsi 4,4,4 simple. Antennae clubbed or expanded filiform. Oval or slightly elongate beetles, finely punctured with flattened pubescence. Associated with fungi, particularly those growing on trees, and with rotting vegetation, litter, etc. Keys out at H26, K43 & K45. Illustrated by Joy, Plate 166 (3-10) and by Lyneborg, Plate 247.

NEMONYCHIDAE. 1 g. 1 sp. 3.0-5.5 mm. Tarsi 4,4,4 with some bilobed segments. Beetles with long rostrum, widened at the front; antennae expanded but not elbowed. Pitchy black beetles with orange coloured hairs, found on conifers. Keys out at G16. Illustrated by Joy, Plate 45 (1).

NITIDULIDAE. 18 g. 96 spp. 1.5-6.5 mm. Tarsi 4,4,4 usually with bilobed segments. Small beetles usually with truncate elytra and very strongly clubbed antennae. Found frequently on flowers, but also on carrion, sap of trees and fungi. Keys out at J16 & K39. Illustrated by Joy, Plate 154-159 and Lyneborg, Plate 231-236.

NOTERIDAE. 1 g. 2 spp. 3.5-4.5 mm. Tarsi 5,5,5. Hind tarsi with swimming hairs. Boat-shaped water beetles, yellowish-brown or reddish-brown, glossy. Keys out at F9. RES Handbook by Balfour-Browne (1953). Illustrated by Joy, Plate 78 (15) and Lyneborg, Plate 49.

OEDEMERIDAE. 5 g. 8 spp. 5.0-15.0 mm. Tarsi 5,5,4 with some bilobed segments. Antennae filiform. Elongate beetles with soft elytra, not closing on centre-line of abdomen in some species, but always with raised longitudinal ridges. Pollen feeders on flowers but also found under bark. Keys out at B9 & G15. RES Handbook by Buck (1954). Illustrated by Joy, Plate 87 (3-5) and Lyneborg, Plate 292-294.

PELTIDAE. 2 g. 2 spp. 5-7 mm. Tarsi 4,4,4 simple to 5,5,5 with some segments bilobed. Antennae clubbed. Oval, flattened beetles with a clear flattened margin to thorax and elytra. Found under bark. Keys out at J4 and K33. Illustrated by Joy, Plate 163 (1).

PHALACRIDAE. 3 g. 16 spp. 1.2-3.5 mm. Tarsi 4,4,4 with some bilobed segments. Antennae clubbed. Oval, convex beetles, usually with distinct sutural striae at the rear. Adults found on flowers, and on fungi. Keys out at J17. RES Handbook by Thompson (1958). Illustrated by Joy, Plate 153.

PHLOIOPHILIDAE. 1 g. 1 sp. 2.3-2.6 mm. Tarsi 5,5,5 simple. Antennae clubbed. Oblong, convex and slightly pubescent beetles, strongly punctured. Thorax dark and elytra with some darker markings. Found in lichen-covered branches of oak trees. Keys out at K20. Illustrated by Joy, Plate 166 (2).

PLATYPODIDAE. 1 g. 2 spp. 4.9-5.5 mm. Tarsi 5,5,5 simple. Antennae clubbed. Cylindrical beetles with very short, clubbed antennae with a scape. Tarsi unusually long. Larvae feed in oak, beech and ash. Keys out at K16. RES Handbook by Duffy (1953). Illustrated by Joy, Plate 66 (1).

PSELAPHIDAE. 19 g. 51 spp. 1-3 mm. Elytra short and truncate, antennae clubbed and often quite long palps. Most species live in leaf litter or in moss. Also known from ants' nests. Keys out at C2 & C7. RES Handbook by Pearce (1957). Illustrated by Joy, Plate 43-44 and Lyneborg, Plate 85-86.

PSEPHENIDAE. 1 g. 1 sp. 1.5-2 mm. Tarsi 5,5,5 simple. Antennae serrate. Small, domed beetles, dark, with deeply furrowed elytra. Found in moist places, sometimes in water. Keys out at H9. Illustrated by Joy, Plate 122 (3).

PTILIIDAE. 18 g. 90 spp. 0.4-0.8 mm. Tarsi simple. Antennae clubbed. Very tiny beetles, with unusual clubbed antennae, living under bark, in vegetable matter, in hollow trees and birds' nests. Keys out at C1 & K2. Illustrated by Joy, Plate 167-168 and Lyneborg, Plate 73.

PTINIDAE. 7 g. 21 spp. 2-5.5 mm. Tarsi 5,5,5 simple. Antennae filiform. Small brown to black beetles, sometimes with rounded elytra and long antennae and legs. Known as spider beetles. Found in dry vegetable matter and stored foodstuffs. Keys out at H16. Illustrated by Joy, Plate 130 (2, 3, 5-11) and Lyneborg, Plate 228-229.

PYROCHROIDAE. 2 g. 3 spp. 7.0-19.0 mm. Tarsi 5,5,4 with some bilobed segments. Antennae serrate or pectinate. Quite large beetles with flat body and elytra broader at rear. Adults found on vegetation, larvae under bark. Keys out at E3 & G10. RES Handbook by Buck (1954). Illustrated by Joy, Plate 87 (1) and Lyneborg, Plate 284.

PYTHIDAE. 1 g. 1 sp. 11.0-14.0 mm. Tarsi 5,5,4 simple. Antennae filiform. Body of beetle very flat; elytra iridescent greenish to brownish, broadest at rear. Keys out at H21. RES Handbook by Buck (1954). Illustrated by Lyneborg, Plate 295.

RHIPIPHORIDAE. 1 g. 1 sp. 10.0-12.0 mm. Tarsi 5,5,4 simple. Antennae pectinate. Elongate beetles with elytra pointed, not meeting on the centre-line of abdomen. Larvae parasitic. Keys out at B5. RES Handbook by Buck (1954). Illustrated by Joy, Plate 89 (1) and Lyneborg, Plate 283.

RHIZOPHAGIDAE. 3 g. 21 spp. 1.5-6 mm. Tarsi 5,5,5 or 5,5,4 simple (at least on hind legs). Antennae clubbed. Elytra truncate, as in Nitidulidae, but tarsi simple and beetles generally more elongate. Associated with tree bark and rotting vegetation. Keys out at K19 & K27. RES Handbook by Peacock (1977). Illustrated by Joy, Plate 142 (13-19) & Plate 148 (9-16) and Lyneborg, Plate 237.

SALPINGIDAE. 5 g. 10 spp. 1.5-4.5 mm. Tarsi 5,5,4 hind tarsi simple. Antennae filiform. Small beetles, some of which have a distinct rostrum. Shining beetles, found under bark or in hedges. Keys out at H20. RES Handbook by Buck (1954). Illustrated by Joy, Plate 87 (6-8).

SCAPHIDIIDAE. 3 g. 5 spp. 1.7-6 mm. Tarsi 5,5,5 simple. Antennae filiform or weakly clubbed. Pointed-oval beetles, with somewhat truncate elytra, found in fungi or rotting timber. Keys out at H5 & K13. Illustrated by Joy, Plate 134 (11-12) and Lyneborg, Plate 99.

SCARABAEIDAE. 21 g. 81 spp. 2.6-20 mm. Tarsi 5,5,5 simple. Antennae with lamellate club. Often large, convex beetles. Found on dung (dung beetles), on vegetation, or flying (chafers). Keys out at D4. RES Handbook by Britton (1956). Illustrated by Joy, Plate 70 (1-4), Plate 71-73 and Lyneborg, Plate 113-145.

SCIRTIDAE. 6 g. 16 spp. 1.5-4 mm. Tarsi 5,5,5 with some bilobed segments. Antennae filiform. Small, yellow-brown beetles with soft elytra, found in damp places. Keys out at G9. Illustrated by Joy, Plate 122 (4-9) and Lyneborg, Plate 147-148.

SCOLYTIDAE. 27 g. 57 spp. 1.2-6 mm. Tarsi 4,4,4 with some bilobed segments. Small, cylindrical beetles with elbowed, clubbed antennae but without a rostrum. Known as bark beetles, some are serious forest pests. Keys out at J10. RES Handbook by Duffy (1953). Illustrated by Joy, Plate 66 (2-13), 67 & 68 (1-6) and Lyneborg, Plate 433-440.

SCRAPTIIDAE. 2 g. 17 spp. 2.0-5.0 mm. Tarsi 4,4,4 simple or 5,5,4 with some bilobed segments. Antennae filiform. *Scraptia* is yellowish with head and antennae dark, and is found in rotting wood. *Anaspis* is variable in colour, but with black antennae with yellow basal segment, and is found in spring on flowers. Keys out at G12 & H23. RES Handbook by Buck (1954). Illustrated by Joy, Plate 89 (2-13).

SCYDMAENIDAE. 8 g 30 spp. 0.7-2.2 mm. Tarsi 5,5,5 simple. Antennae filiform or clubbed. Small, shining beetles, sometimes with oval elytra, clothed in fine, sparse hairs. Found near water, under stones or bark. Keys out at H16, K22 & K23. Illustrated by Joy, Plate 136 (1-12).

SILPHIDAE. 7 g. 23 spp. 9-30 mm. Tarsi 5,5,5 simple (front tarsi may be bilobed). Antennae clubbed or expanded filiform. A very variable group of large beetles, that live on decomposing animal matter and on fungi. Head much narrower than thorax. Keys out at D2, H7 & K18. Illustrated by Joy, Plate 132 and Lyneborg, Plate 75-84.

SILVANIDAE. 8 g. 10 spp. 2-7 mm. Tarsi 4,4,4 with some bilobed segments. Antennae clubbed, sometimes weakly. A rather variable group, including a number of cereal pests. Keys out at J18. Illustrated by Joy, Plate 143 (1-3) and Lyneborg, Plate 238.

SPHAERIIDAE. 1 g. 1 sp. 0.7 mm. Tarsi simple. Antennae clubbed. A very tiny, shining, hairless beetle found under stones, near water. Keys out at K5. Illustrated by Joy, Plate 136 (2).

SPHAERITIDAE. 1 g. 1 sp. 5.5-6.5 mm. Tarsi 5,5,5 simple. Antennae clubbed. Beetles very much resembling Histeridae, but with different arrangement of punctures on elytra. Found in Scotland in coniferous woodland, usually on fungi. Keys out at K12. RES Handbook by Halstead (1963). Illustrated by Joy, Plate 136 (16).

SPHINDIDAE. 2 g. 2 spp. 1.3-2.4 mm. Tarsi 4,4,4 simple. Antennae clubbed. Small black or brown beetles with parts of antennae and legs red. Found in powdery fungi on trees. Keys out at K35. RES Handbook by Pope (1953). Illustrated by Joy, Plate 162 (2).

STAPHYLINIDAE. 188 g. 975 spp. 0.7-24 mm. An enormous family, all of which have truncate elytra exposing at least 3 abdominal segments. Sometimes the elytra are very short, but underneath they have fully developed wings. This family includes the Devil's Coach-horse beetle. Very varied habits, but often associated with decaying vegetable matter. Keys out at C7. RES Handbook (in part) by Tottenham (1954). Illustrated by Joy, Plate 4-42 and Lyneborg, Plate 87-98.

TENEBRIONIDAE. 30 g. 44 spp. 1.5-25.0 mm. Tarsi 5,5,4 usually simple (front tarsi may have bilobed segments). Antennae filiform or clubbed. A group of beetles associated with dry places. Dark, often with sculpturing on thorax or elytra. The larvae include meal worms. Keys out at H19 & K30. RES Handbook by Brendel (1975). Illustrated by Joy, Plate 90-91 and Lyneborg, Plate 265-274.

TETRATOMIDAE. 1 g. 3 spp. 3.0-4.5 mm. Tarsi 5,5,4 simple. Antennae clubbed. Rather cylindrical beetles, shiny, punctured. Found on fungi on wood. Keys out at K28. RES Handbook by Buck (1954). Illustrated by Joy, Plate 92 (2).

THROSCIDAE. 1 g. 5 spp. 1.5-3.3 mm. Tarsi 5,5,5 simple. Antennae clubbed. Beetles related to Elateridae, but with clubbed antennae. Found in sandy areas. Keys out at K21. Illustrated by Joy, Plate 127 (1).

TROGIDAE. 1 g. 3 spp. 5-12 mm. Tarsi 5,5,5 simple. Antennae with lamellate club. Broad, convex beetles with obvious sculpturing of the thorax and dull elytra. Found in dry carrion, old bones and birds' nests. Related to the Scarabaeidae. Keys out at D3. RES Handbook by Britton (1956). Illustrated by Joy, Plate 70 (8) and Lyneborg, Plate 106.

TROGOSSITIDAE. 2 g. 2 spp. 5-11 mm. Tarsi 4,4,4 simple. Antennae clubbed or expanded. Associated with stored cereals, rotting timber and fungi. Antennae with only 10 segments. Keys out at H25 & K40. Illustrated by Joy, Plate 137 (8) & Plate 148 (4) and Lyneborg, Plate 214.

References

Note. SBF refers to the *Synopses of the British Fauna* series published by Field Studies Council on behalf of the Linnean Society of London. RES refers to the *Handbooks for the Identification of British Insects* series published by Field Studies Council on behalf of the Royal Entomological Society. Volumes in both series may be obtained from Field Studies Council.

Balfour-Browne, F. (1953). *Coleoptera: Hydradephaga*. Handbooks for the Identification of British Insects, Vol. IV Part 3. Royal Entomological Society of London.

Brendell, M. J. D. (1975). *Coleoptera: Tenebrionidae*. Handbooks for the Identification of British Insects, Vol. V Part 10. Royal Entomological Society of London.

Britton, E. B. (1956). *Coleoptera: Scarabaeoidea*. Handbooks for the Identification of British Insects, Vol. V Part 11. Royal Entomological Society of London.

Buck, F. D. (1954). *Coleoptera: Lagriidae to Meloidae*. Handbooks for the Identification of British Insects, Vol. V Part 9. Royal Entomological Society of London.

Clarke, R. O. S. (1973). *Coleoptera: Heteroceridae*. Handbooks for the Identification of British Insects, Vol. V Part 2(c). Royal Entomological Society of London.

Crowson, R. A. (1981). *The Biology of the Coleoptera*. Academic Press, London.

Duffy, E. A. J. (1952). *Coleoptera: Cerambycidae*. Handbooks for the Identification of British Insects Vol. V, Part 12. Royal Entomological Society of London.

Duffy, E. A. J. (1953). *Coleoptera: Scolytidae and Platypodidae*. Handbooks for the Identification of British Insects, Vol. V Part 15. Royal Entomological Society of London.

Freude, H., Harde, K. W. and Lhose, G. A. (1965). *Die Kafer Mittleuropeas*. Goecke and Evers. Krefeld.

Halstead, D. G. H. (1963). *Coleoptera: Sphaeritidae and Histeridae*. Handbooks for the Identification of British Insects, Vol. IV Part 10. Royal Entomological Society of London.

Johnson, C. (1966). *Coleoptera: Clambidae*. Handbooks for the Identification of British Insects, Vol. IV Part 6(a). Royal Entomological Society of London.

Joy, N. H. (1932). *A Practical Handbook of British Beetles*. Reprinted (1976) in reduced format by E. W. Classey Ltd., Farringdon, Oxon.

Kerrich, G. J., Hawksworth, D. L. and Simms, R. W. (1978). *Key works to the Fauna and Flora of the British Isles and Northwestern Europe*. Systematics Association Special Volume, No.9. Academic Press, London.

Levy, B. (1977). *Coleoptera: Buprestidae*. Handbooks for the Identification of British Insects Vol. V Part 1(b). Royal Entomological Society of London.

Lindroth, C. H. (1974). *Coleoptera: Carabidae*. Handbooks for the Identification of British Insects Vol. IV Part 2. Royal Entomological Society of London.

Lyneborg, L. (1977). *Beetles in Colour*. Blandford Press, Poole, Dorset.

Peacock, E. A. (1977). *Coleoptera: Rhizophagidae*. Handbooks for the Identification of British Insects, Vol. V Part 5(a). Royal Entomological Society of London.

Pearce, E. J. (1957). *Coleoptera: Pselaphidae*. Handbooks for the Identification of British Insects, Vol. IV Part 9. Royal Entomological Society. of London.

Pope, R. D. (1953). *Coleoptera: Coccinellidae and Sphindidae*. Handbooks for the Identification of British Insects Vol. V Part 7. Royal Entomological Society of London.

Pope, R. D. (1977). Kloet & Hincks' *A Check List of British Insects*. 2nd. Edition (completely revised) *Coleoptera* and *Strepsiptera*. Handbooks for the Identification of British Insects. Vol. XI Part 3. Royal Entomological Society of London.

Thompson, R. T. (1958). *Coleoptera: Phalacridae*. Handbooks for the Identification of British Insects, Vol. V Part 5(b). Royal Entomological Society of London.

Tottenham, C. E. (1954). *Coleoptera: Staphylinidae (part)*. Handbooks for the Identification of British Insects, Vol. IV Part 8(a). Royal Entomological Society of London.

Walsh, G. B. and Dibb, J. R. (1974). *A Coleopterist's Handbook*. 2nd. edition (revised). The Amateur Entomologist's Society, Hanworth, Middlesex.